# After the Music Stops
## Released but not Recovered

By Gloria Gayden Corona

# After the Music Stops
Released but not Recovered

**After the Music Stops:** *Released but not Recovered*

Copyright © 2015 by Gloria Gayden Corona

All rights reserved. No portion of this publication may be reproduced, stored in a retrieval system, or transmitted by any means—electronic, mechanical, photocopying, recording, or any other—except for brief quotations in printed reviews, without the prior written permission of the publisher.

All photographs, news clippings, and transcripts of news articles are the property of their respective owners and publications unless otherwise noted.

*Editors:* Kimberly Rooks, Kayte Middleton, Christian Pacheco
*Cover Design:* Jason Kauffmann / Firelight Interactive / firelightinteractive.com
*Interior Design:* Kyle Weichman / Rick Soldin

**Indigo River Publishing**
3 West Garden Street Ste. 352
Pensacola, FL 32502
www.indigoriverpublishing.com

Ordering Information:
Quantity sales: Special discounts are available on quantity purchases by corporations, associations, and others. For details, contact the publisher at the address above.

Orders by U.S. trade bookstores and wholesalers: Please contact the publisher at the address above.

Printed in the United States of America

Library of Congress Control Number: 2015951502

ISBN: 978-0-9962330-2-6

First Edition

*With Indigo River Publishing, you can always expect great books, strong voices, and meaningful messages. Most importantly, you'll always find... words worth reading.*

This book is dedicated to American veterans who suffer from Post-Traumatic Stress Disorder.

It is for us, the American people, that they suffer.

# Contents

*Acknowledgements* . . . . . . . . . . . . . *ix*
*Preface*. . . . . . . . . . . . . . . . . . . . *xi*

Chapters 1–11 . . . . . . . . . . . . 1–129

*Epilogue*. . . . . . . . . . . . . . . . . . *141*
*Afterword* . . . . . . . . . . . . . . . . . *147*

# Acknowledgements

I feel I owe the most gratitude to all the returned American POWs who I met and subsequently admired along the way and who were the catalyst for this book.

I also must thank Indigo River Publishing—Adam Tillinghast and Kimberly Rooks particularly.

Thanks also to Brandie Simmons, Dr. Leo Kling, Lisa Box, Mary Aagesen, and Karen Newman.

Last but not least, I am grateful to my husband, Dr. Barry Corona, who encouraged me to tell this story about the ravages of war and the sneakiness of PTSD.

# Preface

U.S. Air Force Captain Ben Ringsdorf spent six and a half years in North Vietnam as a prisoner of war. This is the story of his imprisonment, his repatriation, and his hero's welcome home to the United States of America. This is the story of how he met his dream family, how he wished to dedicate his life to helping others, and how the horrors of war eventually took its toll on his life and his family. This is the story of an ordinary American who was required to do extraordinary things.

Overall, this book covers his experience of the Vietnam War from Ben's shoot-down date of November 11, 1966 until 1973. Because of his imprisonment in Vietnam, he suffered mentally and physically from what now is commonly known as Post Traumatic Stress Disorder (PTSD).

This syndrome usually consists of combat-related nightmares, anxiety, anger, depression, and alcohol and/or drug dependence. These symptomatic recurrences can be triggered by anything that reminds the soldiers of the stresses of their combat experiences. If a Vietnam veteran had some trouble adapting to civilian life and many years later he experienced life-changing events, it might trigger some PTSD symptoms from long ago. This seems to be unknown to the general public as many people say that they expect a "shell-shocked" soldier to come home from war, not someone in whom a disorder "festered" for many years before exploding forth in the everyday life of an individual. There is growing interest in understanding the interpersonal nature of PTSD for prevention and treatment. This emphasis is especially timely because of current U.S. military operations. It is important to understand this trauma's effect on intimate relationships because 50% of our present service personnel are married and many more are in other committed intimate relationships.

I hope that this book will in some small way help those who suffer from PTSD and their families who are, alongside their loved ones, suffering in their own way.

# One

From the Agence France-Presse:

> *February 12, 1973*
>
> *HANOI, NORTH VIETNAM — Buses, still bearing the camouflage paint of war, arrived at the civilian airport at Gia Lam this morning carrying American prisoners of war in their last moments of captivity. As the hour of release approached, the airport was swarming with more soldiers of varied rank and nationality than anyone here could remember. The formalities were simple and swift. In a little more than an hour, all 116 men were headed by air for Clark Air Base in the Philippines.*
>
> *"They were released as rapidly as they were captured," one North Vietnamese official remarked with a*

*smile. Scores of North Vietnamese officials had left their ministries to cross the Red River to the airport for what all present clearly regarded as an historic moment.*

*There were North Vietnamese and American officers and enlisted men; Canadians, Hungarians, Indonesians, and Poles from the International Commission of Control and Supervision and some 120 Vietnamese and foreign journalists—none, however, from the United States. Also present was a team from the Four-Party joint Military Commission, formed by the United States, North Vietnam, the Saigon Government, and the Vietcong, but Saigon's representatives were reported missing. The public, however, was not admitted. The atmosphere at the airport, which was still scarred from the bombing of last December, seemed somewhat stiff at first but rapidly became relaxed. North Vietnamese soldiers invited the crew of one of the American medical evacuation planes to tea in a building whose windows had been shattered. In turn, a group of North Vietnamese clustered around an American jeep to study the functioning of the radio with which it was equipped. It was 12:30 P.M. when Lt. Col. Nguyen Phuong of North Vietnam*

## After the Music Stops

*presented to Col. James R. Bennett of the United States the first 20 American prisoners. The men had arrived at the airport riding 20 men to each bus. Each group on stepping to the ground was formed into two lines of 10 men each. One by one, the prisoners passed before colonel Phuong and Colonel Bennett, who were seated at a small table, under a canopy of green parachute cloth, installed in case of rain. The table was in a grassy enclosure surrounded by a wrought-iron fence.*

*As each prisoner's name was called he would step into the enclosure, give his name, and in a move signifying repatriation, walk past the table. An American serviceman would then escort him about 75 yards to one of three C-141 Star Lifter transports. Some of the released men saluted at the table. Some did not. One displayed a piece of white canvas bearing, in blue, the words, 'God Bless Nixon and the American People.' Twenty-nine of those released today were wounded or ill. Three were carried aboard their plane on stretchers. Two were using crutches. Six of the wounded were said to have been crew members of B-52s shot down in December. A list containing the name, rank, birth date, place of capture and conditions of each of the prisoners was turned over to Colonel Bennett. By 1:45 P.M., all prisoners were airborne, 40 in each of the first two planes and the 36 others in the third. One more aircraft followed the C130 that had brought a medical team, telecommunications specialists, ground crews and the radio jeep.*

The American servicemen—the POWs—came home! The celebration was dubbed "Operation Homecoming." Like the rest of the nation, and probably most of the world, I watched the event on television. The men were getting off the big transport planes at Clark Air Force Base in the Philippines. In a roped-off area near the flight line, a crowd of more than a thousand people cheered. A color guard of the U.S. Air Force in silver helmets stood stiffly at attention. A ramp was wheeled into position up to the door of the first C-141, and a carpet of red nylon plush was unrolled for the men to walk down and into freedom. Nearby was the welcoming committee, which included an ambassador, an

## After the Music Stops

admiral, and an Air Force general, as well as many other dignitaries. As the POWs deplaned, it was surprising and uplifting to see how well they looked in general after their ordeal. They were thin—some limping and their limbs contorted by torture—as they raised their arms in salute, but all-in-all they were beautiful to all of us viewers. They were finally stepping foot on American soil after having been away for so many years. In fact, my family and I had been away from the United States for about six and a half years, the same as Ben and some of his fellow soldiers. My ex-husband, their father, was a Pan American airline pilot flying out of New York, and we had been living in Portugal until we had divorced.

I've learned that life is unpredictable and can slap you in the face with surprises and inexplicable coincidences. As my family adjusted to a new life and new customs in the United States, so were the POWs. In buffet lines, Ben and others commented on the huge amounts and choices of food served and felt the wastefulness of all the leftovers. The POWs loved the miniskirts on all the women and were surprised by how the men looked with their long hair and sideburns. They took note of the bell-bottom jeans and shirts unbuttoned to the males' navels. Music was blaring all around, loud and out of boom boxes. The POWs were shocked at the free-love atmosphere, the "pushiness" of the women, and the availability of drugs everywhere.

I recall an article I had read in the newspaper about some of the adjustments the POWs were faced with. A lot of changes had taken place after Captain Ringsdorf's shoot-down date of November 1966. To quote The Washington Post, "Old enemies like China were new friends, and there was détente with the Soviet Union. The American domestic scene had altered radically. In 1973 there were rising prices, rising taxes, and 'risen' hemlines. Brassieres had been burned; men's hair had been dropped in length to over shirt collars. There had been Watergate, X-rated movies, and anti-war veterans were throwing their medals on the step of the Capitol building while American flags decorated the seats of Levi's." This quote summed up the changes quite well for the POWs.

The adjustment came in many forms for my family, one of which was, too, in the area of food. The children did not like most American food. They were used to only fresh vegetables and freshly baked bread and pastries. They also preferred all kinds of fish, no hamburgers, and home-made potato chips served piping hot, not out of a sealed bag. They liked homemade lemonade, not drinks out of a bottle. They were used to eating lots of fish, not fast food. They were used to speaking Portuguese except when they played with their English-speaking friends. They went to an English school that was affiliated with the Anglican Church of England. When they played with friends, they were invited to the friend's home at a specific time and played in a walled garden, never in the street or in an open yard. They also missed their burro "Charlie Brown" and their favorite spoiler of children, Isaura, who was a tiny Portuguese woman who spoke not a word of English. Since she doted on my baby João (Jay), he learned Portuguese virtually before he spoke English.

Back in the United States, our lives were busy and full of new things, and we were looking forward optimistically to the life in front of us. We also had this in common with the returning fighting men. As my children and I settled down, this shared experience—though to two very different extents—would become even more of an inexplicable coincidence.

One day, Cornelia called me and asked, "How about going on a blind date tonight, a double date with me and George?"

## Gloria Gayden Corona

George Wallace was a prominent Southern politician and the forty-fifth governor of Alabama. He even ran for president in 1968. Aside from his political career and surviving an assassination attempt, George was most famous, or should I say infamous, for his opposition to the Civil Rights Movement—Ben and I did not support his position on the matter. You may remember him standing in front of the doors at the University of Alabama when the college was desegregated. However, he did eventually reverse his stance on segregation.

Cornelia Wallace was his wife and an old friend of mine from our high school days. *"No way!"* I thought immediately. *"Why would I want to go on a blind date when I've had enough of men in general?"* I had been married for many years to my ex-husband, my high school sweetheart. We divorced because he loved airline stewardesses a little too much for what is acceptable when one is a married man.

Then I realized that Cornelia was thinking of my happiness and looking out for me in the same way that she had found her happiness. I knew her well enough to know that she was an optimistic person with what she thought was my best interest at heart, so I decided to go on the date.

## After the Music Stops

I said yes, and she said with a giggle, "You can do it for your country."

After accepting Cornelia's invitation to join them on the blind dinner date, I planned to meet the governor and his wife at the Governor's Mansion, as I did not want my children involved with meeting my dates. I also felt it would put an undue strain on Ben Ringsdorf to be forced to make small talk with these young children to whom he was a stranger.

As I drove into the driveway at the Governor's Mansion, I took in the beauty of the grounds with the tall mature trees and a multitude of Southern flowers in full bloom, their fragrance perfuming the air. Gardenias, magnolias, sweet peas, nasturtiums, hydrangeas, and pansies were planted all over the walking paths. Even mint grew everywhere and was constantly in use for the tons of iced tea served from this kitchen. The surrounding homes had been lovingly restored and were stately mansions from another era. I thought of Mrs. Ligon, who owned the Governor's Mansion before the state bought it in order to house Alabama governors and their families. I had played in the home many times, accompanying my father when he made house calls to Mrs.

## Gloria Gayden Corona

Ligon. He was her doctor, and she was too feeble to come to his office for her treatments. I remembered riding up and down in the small elevator and spending many happy hours exploring this huge mansion on my own.

I also visited this house frequently when I was in high school and Governor Jim Folsom had moved into the Governor's Mansion. Cornelia was his niece, and her cousin, Rachel (who was also my friend), was his daughter and lived in the mansion with her father. Since Governor Folsom was a widower, Cornelia's mother, Ruby, managed the mansion for her brother and handled entertaining important guests. The governor had two children to raise on his own, which was also a reason his sister had moved to Montgomery to help him. Therefore, Cornelia, Rachel, and our crowd of high school friends had been in the house for sleepovers, tea parties, tennis games, and just hang outs.

It seems ironic that this same location was now in my life once again. How far and wide life had taken Ben and me for us to end up here at the same moment in time. I felt very moved by this turn of events and its improbability. It is remarkable that Ben and I came together in this particular house where I had played as a child. It's amazing for the fact that an Alabama boy had just arrived home from Southeast Asia and an Alabama girl had just moved back home from Portugal. What are the odds of this happening in a lifetime?

I pulled up to the gatehouse at the Governor's Mansion and stopped at the barrier. A state trooper came out and walked up to my car, checking it out as he approached. He leaned down to my open window and looked inside. I told him my name, and he cleared me to enter the premises. The barrier was raised, and I drove up to the back entrance to the mansion, behind the kitchen wing of the building, as I had become accustomed to. I preferred to enter this way as it was less formal and did not require much of a to-do. This was also the way to the tennis courts where Cornelia and I played most days with all of our friends. The grounds of the mansion were quite extensive, including these tennis courts and plenty of green space in this section of the capital city.

As I entered the Governor's Mansion through the kitchen, I breathed in the tantalizing aromas and observed what a bustling, happy place it was. The kitchen was a gathering place for the mansion's staff. At times, it could include as many as ten or as few as two people. Cooks, maids, gardeners, valets, and sometimes even guards would stop in for a cup of coffee. There was always some hot goodies to be eaten there and an enthusiastic greeting to receive. George's daughter from his previous marriage lived there as well as Cornelia's two sons from her first marriage. They were young, growing, and always hungry, so the kitchen was a busy and popular room for Alabama's first family.

That's where Strawberry was, usually ironing and singing. She was Cornelia's personal assistant, a parolee in the prison work-release program. This particular program was designed for incarcerated prisoners who had been convicted of a passion killing, as the odds were against it ever happening again. Her former husband used to tie her up and beat her twenty years before. She had no money and nowhere to go, so one day she felt she had no way to save herself but to shoot him.

Everyone in the kitchen knew all the gossip, and, as I entered, the topic of the day was the planning of Strawberry's summer wedding to Edward, another parolee. He was serving a life sentence for murdering a man who was beating him with a baseball bat. Edward was the governor's private aide and personal valet. George was confined to a wheelchair thanks to an assassin's bullet, and he needed help

## After the Music Stops

bathing, dressing, and getting in and out of cars. He told us that he was in constant pain with the bullet lodged in his spine in such a way that it could not be surgically removed. Regardless, he was a working governor and therefore went to his office in the capitol building every weekday, needing Edward's strong arm in order to manage his schedule.

After I passed through the kitchen and said my hellos, I walked into the main hall of the entrance foyer looking for Cornelia. I heard her chatty tone and followed her voice to find Cornelia on the enormous winding staircase. She was coming down the stairs with a man in an Air Force uniform on her arm. Both were smiling and carrying on an animated conversation, utterly at ease with one another.

Cornelia and Ben Ringsdorf had grown up and gone to school together in Elba, Alabama up until high school when she had moved to Montgomery. She had known Ben in Elba and had kept him in her prayers for the six and a half years he had been in North Vietnam prison camps. Cornelia had worn Ben's POW bracelet with his name, rank, and shoot-down date the whole time without taking it off even once.

When they reached the bottom of the stairs, Cornelia introduced us, and I realized we were each other's contemporary. Captain Ben Ringsdorf was a handsome man, too thin to be sure, but with an outgoing demeanor. I thought that this might turn out to be a fun as well as an interesting evening in spite of my reluctance at first to go on this date. Ben was also seemingly a typical fighter pilot with a raffish air about him. My first husband had been a fighter pilot in the Air Force before becoming a commercial pilot, so I was familiar with this type of personality in a man and was somewhat drawn to it.

As we walked out of the mansion and to the limousine and waited for George to be lifted into his seat, we began to get to know each other. Cornelia had already filled each of us in on the other's background and past, so we didn't have to get into a lot of the painful parts of our lives at such an early moment. All we needed to do was get acquainted and enjoy each other's company. He told me he was still assigned to Maxwell Air Force Base and would be there for about six weeks for his "debriefing" and medical checkups. Everything was all arranged ahead of time by the Operation Homecoming procedures set in place for the POWs.

We were escorted to the restaurant by plain-clothes security men, the state troopers assigned to protect the governor and an Air Force escort officer for Ben. These men fanned out on the edges of the restaurant's main dining room and tried to be unobtrusive. It was quite intimidating to feel oneself so closely observed. However, conversation flowed easily between the four of us with a joyous spirit of camaraderie. We all wanted the evening to be a pleasant experience for the returning POW who had served his country so well.

I remember that Ben asked me what I did with my time, and I explained that the biggest effort in my life was my two children and my adjustment to relocating back home to the United States. My youngest child did not speak much English, and since he was beginning first grade in an American school, we would need to rush his language skills. My oldest had been in an English school, so she did not have a language gap. As for me, I taught French at a local private school and attended graduate school working on my master's degree in guidance and counseling, as well as updating my teacher's certificate.

Since Ben did not seem troubled or startled in any way about the fact that I had children, I intuited that Cornelia had already filled him in about them when she had told him my background. He said that he probably would have had children if he had not been shot down. It was refreshing to meet such a non-judgmental person, and it made being with this man very comfortable indeed.

As the evening and conversation moved on, Ben's family came up when Cornelia asked him how his elderly parents in Elba were handling the crush of the media and all the publicity about his return home. Ben's father had become a virtual recluse since he had retired from his dentistry practice, and his mother was a retired schoolteacher. They were extremely conservative people, and all the attention was distasteful to them because they had suffered tremendously over the years that Ben was gone. For three years, they had had no word from him and no news about him—he was listed as Missing in Action (MIA)—so they did not know whether he was alive or dead. Since he was not married—only engaged—they were his next of kin and had to handle his affairs: deposit his paychecks, put his convertible on blocks in an attempt to save his automobile, and deal with all the details of his life. They had even tried to get back the diamond engagement ring from his fiancée as she had told them she was marrying another man.

## After the Music Stops

There were two other sons—Ben's brothers: Marshall Ringsdorf, a dentist in Birmingham, Alabama, and Frazer Ringsdorf, an employee of Ligget-Meyers Tobacco Company, living in Izmir, Turkey. Marshall and his wife and children were coming to Elba soon for the Ben Ringsdorf Day that was being planned with a parade, reception, and dedication of a monument. Frazier and his family could not come the distance, and they were contemplating when they could see Ben.

Cornelia then told a funny story about Ben's father, Dr. Ringsdorf, who had been her dentist when she was growing up in Elba. She said she was never afraid to go to his office because he lived up to his reputation as a "painless" dentist. Ben laughed at this and said he had heard this rumor about his father for years. He said that it was too bad that his father was not painless when he disciplined his three rambunctious boys at home. Elba was such a small town that everything that the local children did was noticed and reported, so the parents kept up with their children's every move.

That evening at dinner, George asked Ben about his survival training, about how the Air Force had trained him to survive specifically in the jungles of Southeast Asia if he were shot down. He told us that there was a week-long jungle survival school that he had attended at Clark Air Force Base. The school was geared toward the type of combat missions he would be flying and the type of terrain of the country he would be flying over.

At Clark, Ben had briefings on survival methods, wild animals, and food to be found in the jungle. He was taught the military code of conduct and shown how to fill out the Red Cross form for POWs. There was no specific training in prisoner survival methods, as not much was known about how the North Vietnamese would treat POWs.

He was schooled in the different types of snakes indigenous to the Vietnam area and all the common booby traps being used to capture shot-down pilots. Ben also practiced being picked up by a helicopter and then was abandoned in the jungle for five days. He was allowed to carry with him only his gear that he would have on him if he was forced to eject. There was an Air Force survival expert and a local native guide who instructed him how to travel, take shelter, and prepare jungle plants to eat. The guide cooked rice in a bamboo stick and roasted tubers, which were the jungle equivalent of a sweet potato. He was taught, during IMET

## After the Music Stops

(International Military Education and Training), to spot a water tree and how to cut notches in it for an adequate water source.

The soldiers were then split up into teams and spent 24 hours on their own to practice hiding from the enemy. Ben and another pilot were given an hour head start into the dense jungle, and then the natives would try to find them. They were given three tickets, and whenever they were "caught," they had to give a native a ticket. The native could use the tickets for a rice supply. The natives found them hiding right away, as they knew the land better than the back of their hand. When night fell, the dew was so heavy it sounded and felt like it was raining all night. They attempted to sleep on the hard ground even though rats kept running over their faces and bodies. The next morning, they made it to an open field where they saw a helicopter in the air above them, and they shot off their flares and were "rescued."

Ben told us that had previously attended three other Air Force survival schools. One was a week's trek in mountains, and one was living a week under POW conditions and deprivations that included simulated torture. Another was a water survival school in Japan, where he floated in a life raft for several days and underwent a helicopter rescue from the water.

Ben did not start any conversations about his prison days. However, he did seem to like discussing the changes back in this country. He thought that the student war protestors were entitled to their opinion of wanting America out of this war and out of Indochina. This was somewhat of a surprise to me, as all I had heard was criticism about the college students and hippie protestors being anti-American.

George asked Ben what he liked best about the country's changes that had come along since he was away. He said that he really liked the short skirts, and we all laughed that this would be his first favorite impression. George assured Ben that he agreed with him one hundred percent.

George and Ben next got into a deep and private conversation about each of their views of America at that moment in time. Cornelia and I just listened, as it was a revelation to us as to what these two disparate men thought and could agree on. As neither of them would wish to be quoted, they both looked around our sitting area to be sure that they could not be overheard.

## After the Music Stops

They agreed that there seemed to be a new freedom of spirit in America—a less inhibited way of expression, both verbally and in appearances, apparent in the general public. Ben said he had begun to pick up on this feeling at Clark, observing men's long hair, women's short skirts, and the loud, blaring music playing everywhere. Both Ben and George said to each other that they felt that this free spirit was a sign of the maturation of people. Americans were beginning to think and analyze things for themselves.

I remember thinking to myself that perhaps the antiwar protestors had a part in pushing the movement to bring the POWs home. This to me exhibited this new "freedom of spirit" that Ben and George were discussing. I continued to think that sometimes young people and students can be freer to express what some of their elders believe but do not feel free to express openly.

Ben told us about a book that had been compiled and printed in an attempt to aid the POWs in their adjustment back to the States. In it were selected news articles from the last seven years as well as comments and pictures of the war protests and the hippies in their new forms of dress and hairstyles. He said some of the things depicted were shocking and, even though they had heard about all of it from the newest shoot-downs, it still was amazing to see it in actual pictures. He quoted the book as saying that there was black power, yellow power, gray power, and gay power. Also, there was communal living and natural foods as well as marijuana to be smoked and drugs to take and free love to embark on. As a huge baseball fan, he was doubly shocked by the Oakland baseball players wearing white shoes, and he also noticed double-knit suits cut in the styles of the 1930s when he shopped at the PX.

Ben Ringsdorf was an enigma who had every right to be bitter about the years of his life that he had lost and yet wasn't. He majored in chemistry while in college at the University of Alabama, and at the time of his graduation, the young men of America were subject to the draft. Thus, Ben signed on for flight school in the United States Air Force. After flight school, he was sent to MacDill Air Force Base in Tampa, Florida, home of the F-4 Phantoms and a Tactical Air Command training base. Ben was then sent to Vietnam for three tours of duty, the third tour being a direct "invitation" from President Johnson for Ben to return to combat missions. He was encouraged to take a regular commission

## After the Music Stops

in the Air Force, which was why he was invited and considered a volunteer. He declined the regular commission as he wanted to get out of the service after the war and to go to medical school. He wanted to help people as life's work—even after all the darkness in his life.

The evening ended on a positive note—we four had had an open and pleasant time together. At this early stage of our relationship, I observed the quiet demeanor and self-effacing humor exhibited by Ben's personality. He had a self-assured quietness about him, and I remember wondering if this was a result of the solitude that he had been subjected to at times during his years of captivity. However, when he spoke, it was apparent by his words that he had thought carefully of his response. And when he initiated the conversation, I noticed that he could be quite verbose and full of life. One such example of his humor was evident in his remark about President Johnson "inviting" him to return to the air part of the Vietnam War for a third tour of duty. He talked about it as if he had been invited to a tea party that he felt he had to attend.

This evening "on the town" with Governor and Mrs. George Wallace was non-alcoholic, as they did not drink cocktails, so Ben and I did not drink either. Therefore, I had no inkling of how alcohol could affect Ben at that time. After all, he had been locked away in a cell for many years with no access to alcoholic beverages.

As we parted, I knew that I would hear from Ben Ringsdorf again—and I did the very next day.

# Two

The day after our first date, Ben called and asked if he could see me again. We went out that evening with only his Air Force escort officer and the officer's wife. This Air Force lieutenant had been assigned to aid Ben in his adjustment period and was a tremendous resource for all the small details of everyday life in America, some of which had changed so much. We went out to dinner at a cafeteria-type restaurant called Morrison's, and Ben was shocked and commented on the abundance of food. He said it was such a jolt that he had a difficult time making choices. We teased him about his selections, as they were as if a child with an empty stomach had filled his tray. He had three desserts, four different kinds of potatoes, no meat, and definitely no broccoli.

However, he did choose some collard greens and said it was at that moment he knew he was home in the South for sure.

During this time, Air Force dentists went to work on Ben's teeth, and Air Force psychiatrists went to work on his head to help him ease his way to a faster-paced, more demanding life back in America. Intelligence officers rummaged around in his brain for whatever vital information was stored there. Those intelligence debriefings were the most demanding part of Ben's day, for he was forced to relive a nightmare. Obviously, six years would take some time to recount.

The first debriefing sessions sought names of soldiers possibly alive but still missing and possibly still in prison somewhere in Southeast Asia. The officers subsequently wanted from him an hour-by-hour, day-by-day, year-by-year accounting of his imprisonment. What happened to him from shoot-down to confession to liberation needed to be documented. There was no form and no checklist, other than that with the names of other prisoners recounted and the name of every person that he had ever seen in his years of imprisonment. The list was to include the name of every person he had ever heard from, thought he had heard from, or thought he had seen. He was asked to tell the names and locations of the prison camps he had served time in and who were the collaborators, if any.

Days turned into weeks, and the routine continued: medical tests, dental treatments, blood tests, urine samples, stool tests, intelligence sessions, all seemingly without end. Ben was asked what kind of military information the North Vietnamese sought. How were their questions phrased? The intelligence officers wanted to know what kind of political questions had been asked and what kind of politicizing had the North Vietnamese done. Did Ben know what kind of military information the North Vietnamese had? What was the treatment of the American POWs and what methods of torture had they used and who had done the torturing? Who was the camp commander?

Between those debriefings, and when we had the time, we continued to see each other. Once, Ben asked me to do him a big favor. His former fiancé was coming to Maxwell to see him, and this was obviously going to be a very tense and difficult meeting. She had made a big deal over the phone about her husband being generous to allow her to visit Ben in Montgomery. She lived in Naples, Florida, which would make it an overnight visit. Her attitude put Ben in defense mode. Ben felt it would make a positive impression on her if he had a girlfriend drop in on them and prevent him from being an object of her pity. He wanted to show her that he moved on. It was like he wanted to say, "Hey, look! I have someone in my life too!"

As I drove out to Maxwell Air Force Base, the memories of my childhood on the base during World War II came wafting as a memory in my head, though back then it was an Army Air Corps base. I stopped at the guard post, as the signs directed. An MP walked up to my car, asked to see my driver's license, and called the number into his headquarters, and then I was cleared to enter the gate because Ben had left my name and license number with the MPs. It was much easier to enter the base now in peacetime than it had been during the war when civilians would not have been able to go on the base.

The smells on the base rekindled scenes from my past. There was a cotton gin right beside the front gate, which always smelled like bacon being cooked, causing me to salivate at this point. My father had volunteered at age 41 to serve in the Army Air Corps during WWII, as doctors were desperately needed. The Pearl Harbor attack had been the deciding factor in his decision. He was an urologist, so he was doubly needed, as the American army in that era was composed of mostly males.

Ben was housed in the bachelor quarters near the flight line and close to the base hospital where he was being checked and, coincidentally, where my father had practiced. On the way to his quarters, I passed the house where my family had lived for four of my most formative years. It was on "Officer Row," a Spanish-style, large, comfortable house with a red-tiled roof. All the buildings on the base were tile-roofed and Spanish-styled. The Officer's Club was nearby—that's where my brother and I were swimming the day peace was declared and WWII ended.

I parked in a place marked for visitors and began to look for the number on the red-tiled roofs of the sprawling number of one-story buildings. These buildings comprised the bachelor quarters, the housing for single officers assigned to Maxwell Air Force Base as well as visiting officers. I found the number I was looking for and knocked on the door, a little nervously. Ben opened the door with a welcoming smile; he knew who it was by the agreed-upon timing. I hoped that my appearance at his quarters would go far in helping him restore his pride in the presence of the woman who had not waited for his return.

She was petite and blonde and seemed very friendly. But she was someone else's wife and still had not returned the engagement ring that Ben had given her before he left for his final tour of duty in Vietnam. There was an awkward silence after the initial introductions, and I felt that the moment had come for me to make my exit—the mission had been accomplished by my merely showing up.

Another time, in our early courtship, a parade was planned in Elba as part of Ben Ringsdorf Day—the event he had mentioned one of his brothers could not attend when we went out on our first date. In the parade, there were to be marching bands, convertibles, ROTC drill teams, dignitaries riding in convertibles with the tops down, and a lot of smiling and waving. Ben would be riding in the last car of the parade, which was to end at the monument erected in his honor on an island in the intersection of several city streets. The monument was 12 feet tall with Ben's name, rank, and shoot-down date and information about his life chiseled in stone down the front of the obelisk. After the dedication ceremony, there

## After the Music Stops

was to be a reception at the Elba Country Club. The brother from Birmingham and his family were coming, and Ben's parents were to ride in the parade as dignitaries. Ben wanted me to come because he wanted me to meet his family. While we both agreed that it would be tasteless for me to ride in the parade as we hardly knew each other, he said it would mean a lot just to know I was there supporting him.

Since Ben had shared his discomfort about being in large crowds, I realized he dreaded the numbers he was to face on Ben Ringsdorf Day. I felt that I could be of some comfort to this man who had been isolated in a prison camp for so many years. Even though the people crowding the parade route and attending the reception were hometown faces that he was familiar with, there would also be a lot of press. The national news had already been reporting that he was dating Cornelia Wallace's "cousin," which of course wasn't true as I was her friend. They had been bird-dogging us in Montgomery frequently.

Cornelia and George were coming to Elba for the event as well, flying down from Montgomery in the state plane. Cornelia had asked me to ride with her mother, Ruby Folsom, for the one-and-a-half-hour drive. Ruby was not only

## Gloria Gayden Corona

Big Jim Folsom's sister, but also quite a character in her own right. She was over six feet tall with coal-black hair and had a personality as large as her size. She had a reputation as a big drinker and ran with the party crowd in Montgomery, where she was very popular. Ben's mother was a patrician, reserved former teacher, so Ben and I knew she would disapprove of my arriving in the company of Ruby Folsom. Frankly, I enjoyed Ruby as a person who lived life on her own terms and was a generous and non-judgmental type of character. She also had a large extended family in Elba, so she had every reason to be coming for this occasion.

 Ben had been told by his Elba friends that the town was going all out to make this a special day. The city was so proud of their hometown hero and wanted to honor him with a joyous celebration. When Ruby and I arrived, we were shocked by the number of people gathering in the town square and along the parade route. There were American flags and banners flying all over, and we could hear a band playing in the distance. Ruby pulled the car up to the curb in front of Ben's parents' house. Ben was standing in the yard in his Air Force uniform surrounded by a small crowd of people. He came toward me with a big, welcoming

## After the Music Stops

smile. After greeting me, he led me to two elderly people and introduced me to his parents as well as his brother Marshall and his wife and children.

After a full and somewhat hectic day, Ruby picked me up at the country club where the reception was still going on, and we headed back to Montgomery. Since Elba was in a dry county, there had been no liquor served, and I guess that Ruby wanted to get back to one of her parties in the capital city. As we drove through the evening and I was alone with my thoughts, I remembered Ben's description of his childhood in Elba. His mother, Mary Emma, was a graduate of Queens' Women's College in Charlotte, North Carolina. It was a strict Presbyterian school whose president, her uncle nicknamed "Pious Willie." She was deeply conservative and quite religious and had converted Ben's father, Warren Marshall Ringsdorf. Together they had raised their three boys with an iron fist—no movies, no swimming in the public pool, not much of anything that the other children their age were allowed to do. This embarrassed Ben and caused him to be ridiculed by other children.

Ben played the trombone in school band and was particularly good. He and a few close friends formed a combo at school band practice and played together in the afternoons for fun. One night, the combo sneaked out of their homes and met at a road house nearby. The band director at the high school had given his okay, and he had assumed that the boys had told their parents. All of the other boys had told their parents, and one of the fathers had even driven the underage boys to the road house, stayed with them, and then drove them back home. Ben was afraid to tell his parents, and when he was caught, he was promptly sent away to a boarding school in South Carolina for his final high school year. It was a Wesleyan Methodist school, which is an ultra-conservative offshoot of the United Methodist Church. It was an extreme evangelical education and caused him to feel out of place and an oddball later at the University of Alabama.

I thought that these overly conservative and narrow-minded people who were Ben's parents might find it difficult to accept a divorced mother of two into their son's life. Ben had assured me that their opinions did not affect his feelings for us. He said he felt robbed of the chance to raise his own children by circumstances. My feelings were proven to be valid by the somewhat cool reception exhibited by Ben's mother, although everyone else was very welcoming toward me on the day of the parade.

As Ben was wrapping up his debriefing at Maxwell, the Operation Homecoming activities began to ramp up, and so did our dates. The reception Cornelia had at the Governor's Mansion for the returning Air Force POWs from the southeast took place. Many were able to come with their dates or wives and looked so handsome in their dress blues. They did not look as physically fit as fighter pilots usually look, but they were beginning to put on weight and look better.

Around the same time, there was to be a parade and reception to honor the POWs in Dallas hosted by Ross Perot, a well-known hawk for the Vietnam War, a benefactor of the returning POWs, and a supporter of their families. He even ran for president as an independent candidate in 1992 and 1996. Ben asked if I would go with him. There was a large commercial jetliner leased by Perot waiting for us all on the tarmac at Maxwell Air Force Base. It was luxuriously outfitted with a lavish buffet of hors d'oeuvres and pastries.

Everyone was in a celebratory mood and looked like their bodies and health were on the mend. It was wondered if ever before in the history of man had such an unusual and diverse group of men been gathered together. These men were officers in the United States military with an IQ level of 135 or more, their education bachelor's degrees or better. Most were fighter pilots, a group whose very nature was comprised of aggressiveness and some measures of defiance and were known to be venturesome people. They had an esprit de corps, and they believed that they had done a pretty good job against their captors and their enemies. I don't know anyone who thinks otherwise, hawk or dove.

On our return from Dallas, the hours in each day began to blend into days of the reality of everyday life. Ben was feeling extremely tired, as were other POWs—they felt physically and mentally worn out. They all were drained by tension and more than six years of continuous anxiety. Hanoi was so distant now, and yet seemed to be always nearby. A jingle of keys, soft footsteps in the dark of night, brought Ben upright in bed, terror-stricken and soaked with sweat.

# Three

At this time, Ben began opening up and sharing some of his most disturbing memories with me. Ben had begun hearing more and more survival stories about his fellow POWs. Some of these stories he had heard in the Hanoi Hilton, and some he was just hearing after he came home. However, he told me about himself mainly.

The Hanoi Hilton, officially known as Hỏa Lò Prison, was a large building built in the 1800s by the occupying French. It was located in the middle of downtown Hanoi. It was dubbed "Hanoi Hilton" by the American GIs and had been used to house political prisoners sometime after its erection. The first American prisoner was housed there in 1964. The Americans not only were imprisoned in Hỏa Lò, but they were frequently interrogated and tortured by the North Vietnamese, who were searching for military information at this location. With the prison right in the middle of a busy city, it would have been impossible for prisoners to escape and hide among thousands of Asians, who were generally much smaller than Americans.

When Ben was shot down, he was an Air Force captain, a jet fighter pilot flying the famed F-4 Phantom. He ejected after the plane spiraled down and after his back-seater had safely ejected. He landed in a rice paddy and was immediately captured. There were several planes shot down that day by a barrage of anti-aircraft fire from the North Vietnamese. Ben was beaten unmercifully with a rifle butt by angry enemy troops, who were enraged at the destruction caused by these fighter jets strafing from above. He was marched through the countryside on the way north to Hanoi, and in each village, the townspeople threw rocks and spit at him, yelling unintelligible words. It seemed as if the march was for the sole purpose of allowing the peasants to beat him with hoes, rakes, and rocks and attempt to tear off his clothes and survival gear.

After the shock had worn off, Ben reached a conclusion. These peasants had been bombed and strafed by United States aircrafts for a long time. Even though he was terribly frightened, he understood their anger. Their family members had been killed, and their homes and crops, destroyed. He probably had just killed their husbands, sons, fathers, wives, sisters, or daughters. So when they came face-to-face with a pilot who had been conducting the bombing, their anger boiled over.

Ben recalled months later in the Hanoi Hilton, after one of his torture sessions, a Viet Cong officer had sat across from him in the interrogation room of the hooches. The officer was called "Dum Dum" by the POWs behind his back, and he was one of the main officers who ordered torture. This officer began lecturing Ben on the immorality, the illegality, and the unjustness of America's participation in this civil war between the North and the South of Vietnam. He also described the evil deeds of the American pilots who killed innocents. Ben said that particular lecture went on for two hours with the same things being said over and over.

However, the enemy military men, who hated the Americans every bit as intensely, knew how important the captured pilots were for information and for barter in the event of future negotiations for a peace accord between the two governments. They didn't allow the villagers to kill the captured men. He remembered one elderly peasant woman was throwing clumps of mud at him persistently, hitting him in the face and all over his body. After what seemed like an eternity, some Viet Cong soldiers reappeared and led him away. It was as if they had set him up and allowed the old woman to vent her anger. The mob of villagers who pelted him with rocks might have killed him otherwise.

He remembered thinking, "I need to survive," and he stiffened his spine and walked as erect as he was able. He spent several nights in hooches on his way to Hanoi, marching in the daytime and being shoved along by rifle blows. He knew there were other American captives behind him, so he did not feel totally alone. The trap of the large number of anti-aircraft guns that day and the suddenness of their capture meant that the rescue helicopters had not been able to reach them in time to pick them up. Now they were on their way to Hanoi and many years of imprisonment. Ben told one group of questioners at a news conference back in the United States that he could hear aircraft overhead and the roar of the anti-aircraft barrages, and he knew his buddies were looking for him. It seemed to add to his feelings of hopelessness.

Another evening, Ben told me about communicating with other POWs. The North Vietnamese fought this communication all of the time in order to interrupt the contact between the prisoners. The guards and camp commanders feared an organized rebellion. If the POWs were caught trying to communicate with any of the other prisoners, they were punished by being made to kneel down on the cement floor, stretch their arms about their heads, and stay in that position for hours, sometimes days. Another type of punishment had the prisoners sit on a stool. If he fell off, they would put him back on and tie him there. One prisoner sat on that stool for 27 days and nights. Sometimes prisoners caught communicating would be beaten with a hose or a fan belt from a car. Some men took several hundred hits with a fan belt, most of them in the face.

There was also the "rope trick." A length of nylon cargo strap was knotted tightly around a prisoner's left arm, just above the elbow, and the other end passed over and around his right arm. He was rolled onto his side, and a Vietnamese guard stood on his arms as he slowly drew the elbows together. More lengths of strapping bound his legs below the knees and at his ankles. One guard seized the end of the strap binding the prisoner's arms and heaved upward. Then the torturer took one end of the arm straps and looped it around the prisoner's neck, and he pulled up on the end from the legs, drawing the heels against his rear end and then tied the two together. A prisoner would be left like that for days.

And yet the greatest morale booster for the POWs was that communication among each other. Ben told me about the comfort and relief he felt when he first arrived at the Hanoi Hilton and heard the first taps on the wall coming from the next room. As each day passed, Ben heard other taps coming from the opposite wall. These taps instructed him on the tap code, and he readily learned it. He said that it was his lifeline. The code had been perfected by the earlier shoot-down victims and the trial and error method that they practiced. It was the most immediate necessity when a new POW arrived. They all knew that the contact and comfort that the communicating offered was the main thing that each man ached for.

## After the Music Stops

The numbers could be rapped, waved, scraped, or even swept with the sounds of a broom. The alphabet was divided into five parts with five letters in each part, using "C" for "K." Each letter would be tapped in two parts. The first tap would indicate which of the five parts of the alphabet, and the next tap would indicate which letter in that part it was.

There was also a code that used sneezes, coughs, and sniffs that was easy to employ since so many POWs had colds and breathing problems. The North Vietnamese had a hard time enforcing the detection of this code. The number of each sound was the tip-off—two coughs or two sniffs were the number two.

I remember Ben telling me, with laughter and a twinkle in his eye, about sweeping with a broom. It did not matter which POW the Vietnamese guards took outside to sweep the courtyard, the prisoner got a message out to his American POWs. The broom was constructed of bamboo shoots, so the swooshing sound got out the tap code quite well. And the guards never had a clue, which was a huge morale booster on its own.

Almost every night, a series of taps were transmitted throughout the building when the 9:00 P.M. gong sounded: G N G B U (Good night, God bless you).

Ben drew this diagram for me as well as for anyone who asked him to show them:

| Taps | Group 1 | Group 2 | Group 3 | Group 4 | Group 5 |
|---|---|---|---|---|---|
| 1 | A | B | C | D | E |
| 2 | F | G | H | I | J |
| 3 | L | M | N | O | P |
| 4 | Q | R | S | T | U |
| 5 | V | W | X | Y | Z |

This chart was what the tap code was based on. For example:

With a lot of practice, this tap code could be transmitted with great speed. All they really had to remember were the letters in the first column down: A, F, L, Q, and V. They would listen for the first tap which would tell them which of these five letters it was (one tap A, two taps F, three taps L, etc.), and then the next taps just continued across through the alphabet starting from the first tapped letter. After doing it for some time, they all got very proficient at it. The hardest part was making words out of the letters. There would be a definite pause after each word and a series of taps after each sentence. They had special signals for danger and for starting and stopping the messages.

# Four

After the many months of debriefing at Maxwell finally ended, Ben was free to come and go as he pleased. Although he was still technically assigned to Maxwell, he was considered on leave. So we began dating on a regular basis and really began to get to know one another. We often would drop by the Governor's Mansion, play a game of tennis, and have lunch with George and Cornelia. We had some in-depth discussions about the Vietnam War and the futility of carrying on what the French colonists had begun. Should America have wasted any more of our men's lives and our resources to keep the south from being overrun by the north with their onslaught of communism? In the end, we four believed it was time to end the bloodshed.

As Ben and I began seeing each other more frequently, my friends and family began sharing with me their fears about our relationship. I remember particularly that my brother, a physician, warned me not to marry Ben. He pointed out that the medical profession did not know the long-term effects of the POWs' imprisonment and their torture. It was not known, the physical extent of the damage or particularly the mental deterioration caused by the terror. At that time, the PTSD diagnosis was in its infancy as a mental condition.

Suicide and violent acts could follow in their adjustment back into civilian life. The National Institute of Mental Health explains PTSD this way: "When in danger, it is natural to feel afraid. This fear triggers many split-second changes in the body to prepare to defend against the danger or to avoid it. This fight-or-flight response is a healthy reaction meant to protect a person from harm. But in post-traumatic stress disorder (PTSD), this reaction is changed or damaged. People who have PTSD may feel stressed or frightened even when they are no longer in danger."

My brother felt it could potentially be a dangerous situation for me and for the children—that it was too risky a situation to take my children into. Would he develop alcoholism or PTSD? My friends were expressing their concerns also, and I began to observe Ben more closely. He seemed to be settling into daily life in the United States remarkably well. He got along with everyone he met, and he was particularly fond of my children.

Ben's family was also concerned about his becoming too serious with a woman too soon after his return. This was as understandable as my family's concern about his mental state. I could understand both points of view, as I was having my own reservations about our growing attachment to each other. Ben and I discussed these issues, and we agreed to take things slowly for the sake of all concerned.

My children, Leslie and Jay, were nine and six years old respectively when we moved back to Montgomery from Portugal. Their father had completely cut off ties with them, and he really was not a presence in their lives (He did not see them until Jay was about 18.). The kids were in need of a father, and Ben said being in prison had robbed him of a chance to have children of his own. He said he had prayed about a family in prison and had realized that when he got home, realistically, he would probably meet a woman with children—a "ready-made" family. He had also been praying about this possibility since his return. More and more, he wanted to include the children into our plans.

As we kept seeing each other, we included the children on many more outings and they started to open up to him in a loving way. Ben had such a quiet demeanor and seemed so calm and patient with the children that I was impressed. I do not know of any woman who does not want a stable home for her children. And Ben's personality coupled with my Southern background seemed almost too perfect. Life became busy and events were moving swiftly. Not only was I spending more time with Ben, but I was also still working on my master's in guidance and counseling and teaching French at a private school in Montgomery.

We took the children to Orlando for a real fun-filled vacation. The Orlando Welcome Back Committee went all out for the POWs. They gave us a free car to use, free lodging and food for a week, and discounted tickets to all the area attractions. We stayed at the Sheraton Catalina and went to Disney World, Cypress Gardens, and Busch Gardens.

When we got home, newspaper articles were popping out all over.

# After the Music Stops

> **Footnotes in the News...**
>
> Montgomery Examiner
>
> April 1973
>
> MONTGOMERY — Not all POWs, who have completed medical checkups and military debriefings, are leaving Maxwell Air Force Base. According to Col. Clay Leyser, head of Operation Homecoming at Maxwell, Elba's Capt. Ben Ringsdorf makes frequent trips back to the bachelor's officer's quarters. "I think he is chasing Mrs. Wallace's (Cornelia) cousin around," the Colonel said.

And so we realized that our budding relationship had not gone undetected. We both had come to understand that our feelings for each other were growing deeper and deeper. Ben had told me repeatedly he had accepted that his fiancé had married while he was gone.

I cannot say that Ben's proposal was romantic, but it was very moving. With tears in his eyes, he told me he had been praying that the Lord would help him find someone else because he really needed someone in his life, and that someone with children would be an added bonus since he wanted very much to be a father. Ben told me he felt that God had heard his prayers and sent me to him.

Ben said, "Gloria, I love you. And I love Jay and Leslie, too. I want all three of you to be my family, and I will always be grateful for you."

I responded, "We all have come to love you too, and we would be happy to be your family. I love you, Ben, and I want you to know that I am tremendously proud of you, and I am honored to be your wife."

How does one say no to a proposal like that?

### Freed POW Planning to Marry

*Associated Press*

*April 1973*

MONTGOMERY — Only ten weeks after his return from 6½ years in North Vietnamese prison camps, an Air Force Captain is planning to marry for the first time.

Captain Ben Ringsdorf and Gloria Gayden Hall, a French teacher at a Montgomery school, were issued a marriage license Tuesday by the Montgomery County Probate Office.

Neither Ringsdorf nor Ms. Hall could be reached for comment about their plans.

Both are friends of Mrs. George C. Wallace, the governor's wife. She was a schoolmate of Ringsdorf's in Elba, Alabama and came to greet him when he returned to Maxwell Air Force Base here on February 21, 1973.

## Ringsdorf to Wed…

*Spring 1973*

MONTGOMERY — *Some former prisoners of war and their wives had a second wedding ceremony after their return to the United States. But Alabama's Air Force Captain Ben Ringsdorf is scheduled to hear wedding bells for the first time.*

*Ringsdorf and Gloria Gayden Hall of Montgomery were issued a marriage license Tuesday from the Montgomery County Probate Office.*

*The state's first lady, Mrs. Cornelia Wallace, is a former classmate of Ringsdorf's and a close friend of the bride-to-be and was on hand to greet the Captain when he arrived in Montgomery on February 21, 1973.*

*Ringsdorf, 33, was a prisoner of war for 6 ½ years in North Vietnam. His F4 Phantom jet was shot down while on a combat mission on November 11, 1966.*

## Wallaces Attend ex-POW's Wedding

*May 5, 1973*

MONTGOMERY — *Former prisoner of war Air Force Captain Benjamin Ringsdorf and Gloria Gayden Hall were married Thursday on May 5, 1973 with Governor and Mrs. George C. Wallace as wedding party guests, Mrs. Wallace was quoted as saying on Friday. The couple was married in a chapel at Maxwell Air Force Base here by Chaplin Dave Kirk and went on a short honeymoon (a longer one to be scheduled at a later date), according to Mrs. Wallace.*

*Mrs. Wallace, a former classmate of Ringsdorf in Elba, Alabama and a friend of the bride in Montgomery and she introduced them on his return to the United States.*

# Five

By the end of March 1973, the remaining American prisoners had left Hanoi. Across the U.S., in more than a dozen separate press conferences, the earlier returnees related the horror stories of their years in prison camp. Only a comparative handful spoke out, but it was estimated that between 1965 and 1969, 95% of the Americans captured by the Vietnamese were tortured or abused.

Some people suspected that our fighting men cracked under the interrogation, but every man has a completely different physical and mental makeup. I remember at Ben's speaking engagements, he was occasionally asked if he or anyone he knew resisted the torture. The implication was,

"Did anyone stand up to the Vietnamese guards?" I really admired Ben for not blowing his stack at this veiled insult.

POWs are military representatives of the American people, so our military men have a code of conduct. This code of conduct was an outgrowth of studies made after the Korean War to strengthen the American fighting man, to let him know what was expected of him, and to give him a code of ethics and honor:

### Code of Conduct

1. I am an American fighting man. I serve in the forces which guard my country and our way of life. I am prepared to give my life in their defense.

2. I will never surrender of my own free will. If in command, I will never surrender my men while they still have the means to resist.

3. If I am captured, I will continue to resist by all means available. I will make every effort to escape and aid others to escape. I will accept

neither parole nor special favors from the enemy.

4. If I become a prisoner of war, I will keep faith with my fellow prisoners. I will give no information nor take part in any action which might be harmful to my comrades. If I am senior, I will take command. If not, I will obey the lawful orders of those appointed over me and will back them up in every way.

5. When questioned, should I become a prisoner of war, I am required to give only name, rank, service number and date of birth. I will evade answering further questions to the utmost of my ability. I will make no oral or written statements disloyal to my country and its allies or harmful to their cause.

6. I will never forget that I am an American fighting man, responsible for my actions and dedicated to the principles which made our country free. I will trust in my God and in the United States of America.

It is by the code of conduct that each and every POW would like to perform to perfection, to be able to withstand an unlimited amount of pain, and most American POWs tried to the utmost of their ability to live up to that code. This included being subjected to extremely painful and enduring torture. Many of those who did not come home were tortured to death trying to live up to the code. But some people cannot take that amount of pain.

In their debriefings back in the United States, the POWs said that they did not know of a case where an individual did not give something up. That something may or may not have been the absolute truth, but they knew that everyone gave at least something through torture. This was the POWs' way of resisting; when all else failed, they gave little tidbits of false information. Sometimes, this was detected, and it made matters worse for the POW. For instance, when one pilot was asked the name of his backseater, he said "Mickey Mouse." That answer got him a terrible torture session.

Ben had told me on another occasion that anything strategic he had known on the day of his capture was changed that day. He knew it would be changed so his information to the North Vietnamese would be useless. He said that was one of the reasons torture was so futile because anything of value to the enemy—locations, call numbers, bases, missions—was immediately obsolete the moment the pilot was captured.

Even so, giving any information greatly disturbed Ben, as well as all the others reportedly, and they individually were mortified. The feeling that one betrayed his country was common. Even though they suffered extreme pain and, through torture, gave something up, they did recover mentally and physically and would start their resistance posture all over again.

At the Red River Rats convention in Las Vegas that I went with Ben to later on, I heard this story about a Navy pilot who was being tortured for information. He did not know the answers to the particular questions because they included erroneous facts, and there were no answers. The Vietnamese interrogators wanted him to tell the names of all the pilots on his aircraft carrier who had refused to fly combat missions because of their opposition to the war. As this had never happened, the prisoner had no names to give them. The Vietnamese did not believe him and were increasing the torture. So the POW decided to give them names: Captain Marvel, Captain Dick Tracy, and Dr. Ben Casey. The interrogators believed him and publicized these names over the world at a big anti-war conference in Japan and at the International War Crimes Tribunal in Sweden. When the hoax was discovered, the POW was put in irons and solitary confinement for several years.

Ben was told about a new survival training center at Fort Belvoir, Virginia, for young Army lieutenants and focused on what one should do should he be captured. This camp incorporated what was learned by the Vietnam experience. Instead of open defiance of captors, the lieutenants were taught to develop more subtle forms of resistance. At the training camp, those who chose to answer interrogators belligerently were put into a dark room, actually a 2½' × 6' locker buried in the ground, for ten minutes.

In similar situations in 1967 Vietnam, the confinement lasted a good deal longer—over a month. What made the dark even more terrifying were the attacks from cockroaches and rats fighting the prisoner for his bread ration. He learned that when he heard the rats coming, he should throw a crumb away from his body so the rats would leave

## After the Music Stops

him alone. It was a different story with the roaches, which would crawl all over his face as he tried to eat the bread. He would end up eating breadcrumbs along with cockroaches.

One prisoner told of how the heat and darkness caused his whole body to itch. He could not shave or trim his beard in pitch-black dark, and his hair grew very long. After about three weeks, he felt he was losing his sanity. He was afraid they were going to torture him again and that they would break him. In his time of deepest despair, he heard thumps and the sounds of the tap code. He knew then that another POW was in the adjoining cell. The code was his lifeline to sanity, the realization that others had gone through what he was going through and that they had survived. That particular POW survived 48 days of that confinement.

Later, as Ben and I traveled around the country, it was interesting, and sometimes amusing, to hear the questions posed or comments made to him about his time as a POW. One man in an airport recognized Ben from his release pictures in the newspapers and came up to him to ask him a question. He wanted to know why Ben did not try to escape from prison in the big city of Hanoi. Ben, very calmly, turned the question around and asked the man where a person—especially a Caucasian or African American—could hide when he was double the size of everyone else in the entire Asian country.

Ben had another story to share with groups back in the United States when he was on a speaking engagement when asked about escaping. It happened in May 1969 at the camp known as "The Zoo." Two Air Force captains had been planning an escape for more than a year. However, the two fliers were quickly recaptured and brutally punished. One of them survived, and the other was never seen again. The escape provoked wide controversy because it set off a period of retribution, the most brutal wave of tortures and punishments of the war. Every camp was shaken down, every cell was inspected, and almost every prisoner was beaten or questioned. Carefully built communication systems were destroyed. So if escaping meant having other POWs tortured, maybe it was best not to chance it.

No matter where we were or to what group of people Ben was giving a pre-arranged talk, one point that kept coming up was about the Geneva Convention. In general, most people believed that the POWs could not be tortured due to the Geneva Convention. Ben always answered this inquiry

## After the Music Stops

by telling the following story. One time, on the journey to Hanoi, through hostile villages, his captors put him in a room in a hooch. One of the Vietnamese soldiers who spoke passable English started asking for his name, serial number, and rank. According to the code of conduct, that is the only information Ben was required to give to the enemy authorities. The interrogator also asked him questions about his aircraft, the F-4 Phantom jet, his organization, and where it was based. Ben refused to answer, and the questioner became agitated. He began beating Ben and calling him a criminal. He screamed that Ben would be severely punished because he had not shown the proper attitude. Ben told his interrogator that he was not a criminal. He was a prisoner of war and should be treated as such. The Vietnamese officer said that North Vietnam did not recognize Americans as prisoners of war under the Geneva Convention because the United States Congress had not officially declared war. To them, the Americans were criminals and would be treated as such, and since this criminal (Ben) had not shown the proper respect, he would be punished.

A photo of Ben and his fellow POWs finally leaving for home; photographer unknown (presumably a military photographer)

Cornelia's reception for POWs at the Governor's Mansion: Governor Wallace, Ben Ringsdorf, Gloria Montgomery, Alabama 1973; photo taken by an official State of Alabama photographer

## After the Music Stops

### Freed POW Planning To Marry

MONTGOMERY, Ala. — Only 10 weeks after his return from six years in North Vietnamese prison camps, an Air Force captain is planning to marry for the first time.

Capt. Ben Ringsdorf and Mrs. Gloria Gayden Hall, a French teacher at a Montgomery school, were issued a marriage license Tuesday by the Montgomery County Probate Office.

NEITHER RINGSDORF nor Mrs. Hall could be reached for comment on their plans.

Both are friends of Mrs. George C. Wallace, the governor's wife. She was a schoolmate of Ringsdorf's in Elba and came to greet him when he returned to Maxwell Air Force Base here Feb. 21.

### Footnotes IN THE News...

Not all POWs who have completed medical checkups and military debriefings are leaving Maxwell AFB.

According to Col. Clay Leyser, head of Operation Homecoming at Maxwell, Elba's Capt. Ben Ringsdorf makes frequent trips back to the bachelors' officers' quarters.

"I think he's chasing Mrs. Wallace's (Cornelia) cousin around," the colonel said.

+ + +

The football bug is apparently capable of surviving life in a Vietnamese prison camp.

As soon as a formal press conference for POWs was over this morning, Capts. Kevin Cheney and David Dingee, rushed reporters wanting to know the final score of the Super Bowl last January.

"We've been given different scores and we want to know the real one," the two officers said. By the way, the score was Miami 14 and Washington 7.

### Ringsdorf Weds; Wallaces Watch

With Gov. and Mrs. George C. Wallace looking on, an Air Force captain who was held by the North Vietnamese for more than five years was married here.

Capt. Ben Ringsdorf 34, of Elba and Gloria Gayden Hall, 34, a Montgomery school teacher, were married Thursday night at a Maxwell Air Force Base chapel.

The bride and bridegroom were both high school classmates of Mrs. 'Wallace,' who was on hand to greet Ringsdorf when he returned here Feb. 21.

*Ala. Journal May 5, 1973 Sat.*

### Wallaces attend ex-POW's wedding

*B'ham Post-Herald*

United Press International

MONTGOMERY — Former prisoner of war Air Force Capt. Benjamin Ringsdorf and Gloria Gayden Hall were married Thursday night with Gov. and Mrs. George C. Wallace as guests, Mrs. Wallace said Friday.

The couple was married in a chapel at Maxwell Air Force Base here by Chaplain Dave Kirk, and went to New Orleans for a honeymoon, Mrs. Wallace said.

Mrs. Wallace, a former classmate of the two at Elba, introduced them, she said.

*Sat. May 5, 1973*

### Ringsdorf To Wed Cornelia's Cousin

By VIRGINIA GIBSON

Some former prisoners of war and their wives had a second wedding ceremony after their return to the U.S. But Alabama's AF Capt. Ben Ringsdorf is scheduled to hear wedding bells for the first time.

Ringsdorf and Mrs. Gloria Gayden Hall of Montgomery were issued a marriage license Tuesday from the Montgomery County probate office. Mrs. Hall is a cousin of the state's first lady, Mrs. Cornelia Wallace.

Mrs. Wallace, who is a former classmate of Ringsdorf's, was on hand to greet the captain when he arrived in Montgomery Feb. 21.

Ringsdorf's parents, Dr. and Mrs. Warren M. Ringsdorf live in Elba. Neither Ringsdorf, who now lives in Montgomery, nor Mrs. Hall could be reached for comment on the upcoming wedding.

Mrs. Hall teaches at Greengate School.

Ringsdorf, 33, was a war prisoner for five years. His aircraft was shot down while on a combat mission over North Vietnam Nov. 11, 1966.

Ringsdorf

*This was erroneous — I am not Cornelia's cousin — just a friend*

Newspaper articles at the time (Spring 1973)

## Gloria Gayden Corona

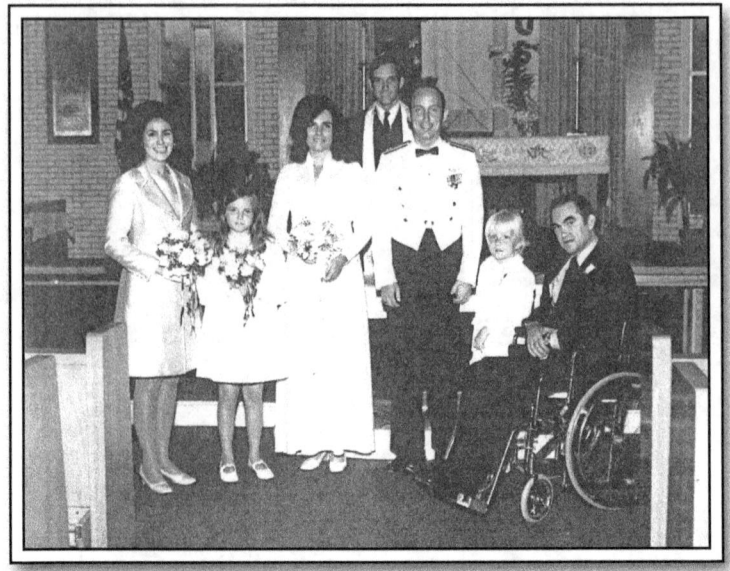

Wedding photo-Maxwell Air Force Base Chapel, Montgomery, Alabama: Cornelia Wallace, Leslie, Gloria, Ben, Jay, George Wallace; photo taken by an official State of Alabama photographer.

Pago Pago, American Samoa: A lush and beautiful port in the Pacific Ocean and a dangerous airplane landing strip

## After the Music Stops

A Fale, a typical Samoan home with grass sides that can be lowered at night

An ocean liner coming into port at Pago Pago

Roofs of the flower-laden tropical hotel complex in Pago Pago

Marist Brothers in Samoa who were building a school

## After the Music Stops

Wedding celebration at Maxwell Air Force Base Officer's Club: Ben, Gloria, and friends

Another picture from the wedding celebration: Gloria, Ben, Col. Don Burns (POW), and Gloria's mother

### Friends Get Together

A number of friends of Capt. and Mrs. Ben Ringsdorf were guests at a cocktail supper when Mrs. Ringsdorf's mother, Mrs. O'Vaughan Gayden entertained in their honor. Talking together at the party are (from left) Mrs. Ringsdorf, Capt. Ringsdorf, Col. Don Burns, and Mrs. Gayden. Both Capt. Ringsdorf and Col. Burns were prisoners of war in Vietnam. See Promenader for details of party.

Ben and daughter Lillian Macon Ringsdorf;
photo taken by a neighbor

Gloria and Lillian; photo taken by a neighbor

# Six

The reason my family was so upset about me marrying a POW was because of thoughts continuously expressed by my physician brother. Needless to say, these warnings still went unheeded, and the wedding had gone on ahead. We married in May 1973 at a chapel on Maxwell Air Force Base.

On May 23, Ben and I flew to Washington, DC, for a dinner at the White House. The POWs had been invited by the president to a reception and dinner with him and the First Lady. The entire affair was to honor the POWs and their service to their country. Ben felt proud to be recognized by his Commander-in-Chief, and I was proud of his bravery and loyalty to his country. We were honored and celebrated from beginning to end.

## Gloria Gayden Corona

When we checked in at the Hilton Hotel, they had put up a large sign disclaiming any connection between this hotel chain and the Hanoi Hilton. There was a breakfast for the POWs, and during the day, the president met with each of us personally and shook our hands. The wives met with Mrs. Nixon for a tea reception. That evening, we had an open tour of all three floors of the White House and then had a scrumptious dinner under a gigantic tent on the White House lawn. Many celebrities and VIPs were there, but they treated the POWs as the important guests that night. Everything was tastefully done, and we had a memorable evening.

Ben and I soon began to make plans for the future, not that we had not discussed our dreams and aspirations before our wedding, but now we found it necessary to get specific. While in prison, Ben began to feel that he wanted to make his life a life of service. With an undergraduate degree in chemistry and an interest in science, the obvious choice was to go to medical school. Ben had early on opted to get out of the Air Force, even though he could have stayed in and let the military foot the bill for medical school. But then he would owe them more years of his life in return, and he felt he had given all the years that he cared to give. In order for Ben to go to medical school, he would have to pass the Medical College Admission Test (MEDCAT), and since we knew he needed to take a refresher course to do well, he enrolled for the fall semester at Auburn University at Montgomery (AUM).

Then we discussed a trip Ben had dreamed of and planned on many dreary and fearful nights in the Hanoi Hilton as a prisoner of war, not knowing if he would live through another day, much less see the outside world as a free man again. Since his brother Frazer and his family had been unable to come to Alabama to welcome Ben home, we planned to visit them there during our honeymoon and to stop in Portugal on our way home.

The first stop on our honeymoon was Mexico City. What a large, chaotic, and overcrowded city! We were shocked at the open, visible disparity between the rich and poor. The Aztec influence was evident all in the city—in how the people looked, in the clothing styles, and in the furnishings. And of course, the food was from the Aztec Indian history and combined with a few Spanish touches. Ben enjoyed the bar in our hotel, where the company was convivial and the tequila was chilled to perfection. When he told me "bye bye" as I left to go shopping, I laughingly told him I would definitely take him up on that—"buy buy."

On the last night Ben and I spent in Mexico City, we went for a night on the town to celebrate the first leg of our honeymoon around the world. We went to several nightclubs and wound up spending at least five hours eating and drinking in a cantina, listening to a mariachi band play Mexican music. The time seemed to fly by, and before we knew it, it was dawn. We hailed a taxi and returned to our hotel. Ben could hardly walk and was tottering along, so drunk I did not know if I could get him back to our room. I had never seen him that inebriated and was somewhat shocked, and a little unsettled, at his condition.

When we got in our room and he fell into the wall, I said, "Ben, I think you've had too much to drink." With that, he picked up the small TV set from the table top and threw it at me. He then fell on the floor and passed out. I was so afraid of him waking up and going after me again that I hurriedly packed and headed to the airport for the next flight to Atlanta. I called my mother in Montgomery to pick me up at the airport. When I got in her car, she let me know how horrifying it was for me to come home from my honeymoon trip like this. How was she going to explain my arrival to her friends? After all, I was already divorced once. We rode in silence after that.

She was looking after my children while Ben and I were gone. Jay was at summer camp already, and Leslie was getting her clothes packed to leave the next morning for her girls' camp. Mother and I saw her off and began to discuss my situation. All this time, Ben kept calling and calling, begging to be forgiven and saying it would never happen again.

My mother felt that I should forgive him and move on, pointing out that he had been locked up for six and half years and that his imprisonment was bound to have a negative effect on his ability to metabolize alcohol. She made me feel so sorry for him. Days later, I flew from Atlanta to Los Angeles to Hawaii and on to Samoa to join Ben. I tried to put the whole issue behind me, but I was constantly afraid of Ben when he was drinking from that day forward.

## Gloria Gayden Corona

I next went through Hawaii at a hurried pace and onto Samoa. We had been scheduled to go New Caledonia and Tahiti, but were unable to continue on our itinerary because of an atomic bomb test being done in that region by France. However, American Samoa was a delightful surprise for me. I had not expected such a lush and beautiful island. Seeing the land from the air in the large PAN American 707 passenger jet before landing, it was distinguished by a verdant mountain rising out of the South Pacific seas.

The landing was also spectacular, but in a different way. The extra-short runway looked like it was resting on a ledge on the side of the mountaintop. It appeared like a shelf on the edge of a green spire with water beneath. Adding to the harrowing nature of this event was the sight of another PAN American plane nose down in the water. It had run out of runway on its landing and was now precariously resting at the end of the tarmac with its tail in the air. The wreck had happened some time before our arrival and had not been removed, as it was awaiting excavating equipment to come from the United States—about a sixteen-hour flight away from Samoa. Our pilot took our aircraft around the area, making three passes in order to "set" it down. That landing was the only time I had known fear when flying!

The capital of American Samoa is named Pago Pago and is quite a popular port for South Seas cruise ships. The beautiful water is extremely deep and therefore a very safe port for the gigantic ocean fairing vessels. The architecture on the island was a mix of Polynesian and Colonial. The houses as well as all structures with a domed roof, circular in shape with wooden posts holding up the thatched roof, were called "fales." The sides were open to allow sea breezes to flow through the house. Grass mats on the sides could be lowered for privacy. The Samoan people we saw were uncommonly large, the men as well as the women, with absolutely the most perfect teeth, dazzlingly white. Our hotel could not have been more charming with domed thatched roofs over the rooms so that it nestled into the mountainside and beside the wide, sandy beach of Pago Pago. Fortunately, the sides of the building would not open, so we were able to have air conditioning. The hotel's cocktail lounge was half inside and half outside and covered floor-to-ceiling with grass mats.

It was the center of activity for the tourists and visitors to the island. Ben learned at the bar about a group of Marist brothers who were in the process of building a school with the help of their Samoan students. We were invited by one of the brothers, Brother Edmond, to have lunch with them on a Sunday, a day when they were not working. The festivities lasted from noon until midnight. The pig had been roasted in a pit in the ground and wrapped in banana leaves. We were served poi and many variations of taro and tofu, which were all quite delicious. We drank wine and liquor the whole day and evening while we were eating. It was a dinner party I would never forget and probably would never be able to live through again.

At one point during our trip around the world, Ben complained about having trouble sleeping when there was too much light in the room. He began to laugh uproariously, and I wondered what was so funny. He told me about a time in the Hanoi Hilton when it seemed that every captive in North Vietnam was moved together to the prison. At that time, a fellow POW knew Ben was having trouble sleeping because of the light in the room (It was on day and night.). This friend constructed him an eyeshade made out of a piece of white cloth and even embroidered it complete with two buttons for eyes. Wearing that, he could take a nap and still look as though he were awake, which brought laughs all around.

Ironically, in the jungles of Western Samoa, Ben came upon a water tree that he remembered from his IMET survival training. He showed me how it could be trimmed with the Swiss Army knife he always carried. When you trimmed away the outer bark, water would pour out and quench one's thirst.

Some of the trees were the height of a three-story building and the same width. When we arrived back in Apia, we were surprised to learn that Robert Lewis Stevenson, the Scottish novelist who wrote *Treasure Island*, *Kidnapped*, and *The Strange Case of Dr. Jekyll and Mr. Hyde*, lived and died on the island. We visited his 400-acre estate in Vailima and felt a strange eeriness to think such an accomplished person had died in such a distant and primitive place, however lush and beautiful it was. We returned to American Samoa and knew we would remember the day forever, standing beside the giant banyan trees.

The next leg of our around-the-world trip took us to Fiji, an island country in Melanesia in the South Pacific about 1,000 nautical miles away from New Zealand. The main island of Viti LeVu has the capital of Suva, where most Fijians live. Fiji has a lot of tourism due to a large amount of soft coral reefs, good for scuba diving. Also, there are a number of really good golf courses, and the elite Fijians are addicted to the game. Ben, a good golfer, got in several games and drinks all around in the hotel bar afterward. All the while, I relaxed in the sun and shopped for tropical treasures.

Our next flight would take us to Sydney, Australia. The long flight took us over the entire continent of and right over the outback, which is nothing but reddish clay. We had only booked three days in Sydney, so we had to rush to see enough to feel satisfied. We took a walk on the Sydney Harbour Bridge to the top of the structure, not the "Bridge Climb," but just the pedestrian crossing that has spectacular views of the city center. We visited the Royal Botanic Gardens and sat on a bench carved out of sandstone in the early 1800s. The gardens were extensive and surprisingly bountiful. We were told to take a walk on Bondi Beach with golden sands, the most famous beach in Australia. Then it was back to the hotel bar to finish our stay down under.

We had a time schedule constraint, as we were due to arrive in Izmir, Turkey and Ben's brother's home. To this day, I happily remember an Australian woman that I met at the Sidney airport who had adopted a rescued injured Koala bear. The Australian government would not usually let a Koala bear be adopted, yet this one was crippled so an exception was made. This lady told me about arriving home each day after work, and the bear would wrap his arms and legs around her tightly in a "bear" hug and would not let her go. So she had to cook and do her chores with him wrapped backward around her body. How delightful to come home to that every day. What a joy!

We boarded Singapore Airlines (still ranked the best airline in the world) and buckled up for the eight and half hour flight from Sydney to Republic of Singapore, a city-state. It is a Southeast Asian island country off the southern tip of the Malay Peninsula, the world's fourth-leading financial center, and its port is one of five busiest ports in the world. Singapore was an unknown entity to me, as I had not known to expect that it would be totally a metropolitan area. In other words, there was no countryside—it was totally a city.

The Chinese are so industrious that there were stores and businesses one on top of the other. They are known as the epitome of entrepreneurs. Everyone was busy working and hustling, so the atmosphere was one of a rushing mass of people going about in every direction.

As I moved around the city, it began to occur to me that people were the same all over the world. Everyone is working for a fulfilling and successful life with comfort for their families, enough to eat, and a decent roof over their head. The Chinese woman selling freshly plucked chickens in her tiny store was just the same as I am. She wanted the same things for her children that I wanted for mine—an education and enough to eat.

We stayed at the famous colonial-style Raffles Hotel, a beautiful place with palm gardens all around it. Since we spent many days in Singapore, we were able to have some clothes tailor-made. Ben had a summer-weight suit and several shirts made, and I was able to select my own fabrics to have some dresses made, one long and two short. Large ceiling fans kept the breezes flowing while I enjoyed riding around in rickshaws to do my buying. If one has never tried it, the experience of riding in a hand-buggy with a man pulling it around the streets next to speeding cars and trucks with seemingly no regard to safety will scare you into walking long distances instead.

Ben spent many hours making friends in the cool, tropical Long Bar. People loved to show him their admiration for his service to his country wherever we travelled. Though he spent a lot of time at the bar, I reminded myself that he had not had a drink or celebratory occasion for six and a half years—a long time in darkness.

Our next stop was in Bangkok, Thailand, a place a fellow traveler told me was once known as the "Venice of the East." We checked into our hotel, the Oriental, which was filled with orchids in its luxurious lobby. A plaque noted that Somerset Maugham stayed there when it was the new posh hotel in Siam. It was tempting to never leave the hotel because it had everything the word Siam evoked. It had a gourmet breakfast served overlooking the busy roaring Chao Phraya River, two staff members per guest to fill a guest's every wish, a seafood buffet at lunch, jazz in the evening, flowers everywhere, and a dark, cool, convivial bar scene. Walking around in the city, we noticed that it was such a jarring mix of modern and ancient. Scattered concrete skyscrapers were built right next to traditional wooden homes and temples gleaming with golden Buddha statues. The smells of jasmine and grilled street food filled the air. Really, the odors of the Orient that we had become accustomed to in Singapore were almost pleasantly overpowering.

From Bangkok, we flew to Tehran, Iran, on a giant 747 BOAC (British Overseas Airways Corporation). We had been told to keep a low profile as unrest in Iran, and perhaps a revolution, was brewing against the Shah and his government. Since the United States backed the Shah, Americans were not looked on too kindly in Tehran. As we entered the airport and got off the gigantic jet, we instantly felt the cold, hard stares from the people milling around the lobby. The women were all in chadors, but their eyes were just as unfriendly as the men. I was glad we were only going to be there for two days on our way to Turkey and Ben's brother and family.

We flew on BOAC to Ankara, the capital of Turkey. Turkey looked like the rest of the Middle East from the air—barren, dusty, and dry with a few trees in mountainous areas. We were changing planes in Ankara for our final destination of Izmir (ancient name Smyrna) on the coast of the Aegean Sea. After a brief layover, we were strip-searched by officials before being allowed to board our aircraft. The men were searched in a separate room from the women, and the search included completely disrobing. It was such a scary and intrusive experience that the passengers were totally shocked, quiet, and acquiescent. Then we boarded Türk Hava, the Turkish national airline, and took off. The plane only taxied a short distance and took off straight up it seemed, which caused all carry-on items to fly around the cabin. Then the landing seemed to be straight down too. No circling the airport for Türk Hava!

Landing in Izmir, we were greeted by Frazer, Gemma, Mary, and John-John. It was so moving to see the hugs and tears for Ben, their returning brother and uncle. They were very loving toward me as Ben's new bride. We were given a short tour of the city before going to their home in a condo in a high-rise building overlooking the port. Izmir has a sweeping seaside promenade that seemed to draw a mishmash of strollers and café patrons. There were groups of women in head scarves side-by-side with packs of college girls in short shorts, mustached fishermen, slick-suited businessmen, and shoppers of every description. We were to spend ten days there, and so we settled into our separate bedroom and bath. The condo was modern and had large windows overlooking the coastline and surrounding mountains. The view was breathtaking. We spent our days wandering the broad boulevards, where the glass-fronted buildings and modern shopping centers were dotted alongside traditional red-tiled roofs, 18th century markets, mosques, and churches.

The United States Air Force had a base in Izmir at that time, and they wanted to give Ben a welcome home parade and reception on the base, also including the Turkish Air Force based there. It was a grand and lavish affair with a band and a parade of dignitaries around the base, with Ben riding on the back of the lead convertible. The evening was filled with buffet tables laden with all sorts of Turkish delights and a surprising assortment of American dishes. Since there were predominantly Muslim guests, liquor was not served, as Muslims do not drink alcohol.

One day, we rode in a minibus that Frazer had hired, driven by an American Air Force sergeant, to take us to Ephesus, a large archeological site near Izmir. The grounds were explored entirely on foot, and it took about two hours. All the paths were clearly marked, and each excavation was explained as to its significance. Ephesus was once the trade center of the ancient world and a religious center of early Christianity; for example, the Gospel of John was written there. The temple to the goddess Artemis was almost four times larger than the Parthenon in Athens. According to the New Testament, the Apostle Paul preached in Ephesus and according to legend caused a riot led by silversmiths who crafted shrines to the goddess and feared for their livelihoods and the future of the temple. Mary, Jesus' mother, was purported to have spent the last years of her life in Ephesus, and there was a house nearby where it was thought that she lived, attended after by John, who was instructed by Jesus to look after her.

## Gloria Gayden Corona

Of romantic interest to me was looking down the wide ancient boulevard toward the sea, picturing Anthony and Cleopatra disembarking their barge and walking up the wide avenue to the city. It was also delightful to see a man coming up the avenue in the distance with "ROLL TIDE" on his baseball cap.

Arriving in Rome, we headed straight to our hotel, the Lord Byron. Traveling around the world in six weeks can get exhausting, and we were both beginning to give out physically. I had been to Rome many times and immediately became rejuvenated by the vibes of this city. It is rightly called "The Eternal City" and has always been my favorite place in the world, next to Lisbon, Portugal, of course.

The hotel was a small boutique-style villa located just off the Borghese Gardens. As we strolled around the area, we found we were near Via Veneto, the Spanish Steps, Trevi Fountain, and the Pantheon. After each day out discovering Rome, we would spend the evenings at the peaceful hotel in the lounge bar, sometimes eating in the hotel restaurant with its award-winning menu and Italian wines. We also made it to the Vatican and marveled at the magnificent Sistine Chapel.

## After the Music Stops

The last stop on our whirlwind honeymoon took us to Lisbon and out to the area where I had lived for close to five years—Cascais. It is a coastal town on the Atlantic Ocean on one side and the Tagus River on one side. We checked into the large Estoril Sol Hotel, and Ben immediately ran into friends from his pilot training days who were now commercial airline pilots. They hustled Ben into the hotel bar to celebrate his homecoming.

While I hopped around Cascais, Malveira da Serra, and Estoril catching up with friends and reliving many happy days in this glorious part of the world, Ben seemed more and more occupied with partying with old friends and new ones in the luxurious Estoril Sol bar.

Looking back on our honeymoon, it almost seemed like an in-depth tour of the world's bars, kind of bitterly ironic. Ben would join me on some fun excursions, but for the most part, he spent his time in hotel lounges and bars. His capacity for alcohol increased so much he could keep a buzz on for hours. While he did enjoy numbing his inner demons, he also loved the adulation he found in the bar crowds celebrating his return to the free world.

# Seven

Afterward, many questions were asked of Ben at the speaking engagements he had around the country. One recurring question was about describing the everyday life as a POW. Some POWs were in solitary confinement from three months to a year. Half were in solitary for up to two years, and others, for four years of the war.

After a prisoner was assigned to a particular prison, a guard would take his old clothes, and he gave the POW a special Vietnamese prison uniform. It consisted of two pairs of long pants, which looked like pajama pants, and two pairs of shorts with short shirts. A prisoner was also issued a pair of "Ho Chi Minh" sandals (as POWs called them), which were made of an old tire tread with inner-tube straps. Some were given an old cotton sweater for when it was cold, but it was so worn from use it did nothing to keep a person warm.

The winter's cold was very penetrating, and a prisoner would put on every article of clothing he had and walk back and forth in a feeble attempt to keep warm. However, whenever POWs were allowed outside, they made themselves take a shower and wash their clothes because they knew that to keep healthy, they needed to try to keep themselves clean. When the earlier prisoners were asked what summer would be like, the guards only laughed. The temperature in the summer got up to 115 degrees outside and inside it was even hotter because there was no ventilation.

Prisoners were also allotted a tattered old mosquito net and a very worn bamboo mat, about a sixteenth of an inch thick, for the concrete slabs, which were to be used for beds. These were so hard they had to turn over constantly because of their aching joints.

Each prisoner was given a little tin cup, a toothbrush, some toothpaste, and a small piece of soap. The soap was to last a person three months and was to be used to wash his body as well as his clothes. When the POW was forced to wash the dishes for his cell block, he was told that if he wanted them clean, he would have to use his soap for that also. Obviously, the dishes went without a soaping for months.

As for shaving, a prisoner was given an old used blade that had been re-sharpened many times on a rock. It was an old plastic razor, and the blade did not really fit it. The only water he could use was his one-cup-a-day tin cup full of water. When shaving was finished, the man's face would look like raw hamburger. They were all allowed to shave once a week, and every man in the cell block used the same blade. The toothbrush was the same story. The handle was always breaking off. When the guard noticed that, he fussed at them for not knowing how to use a toothbrush, and he would hold a match under it to weld it back together.

They were given regulations they had to abide by or be punished. The Vietnamese desired that the POWs show them proper respect. One way they cooked up to do this was to make the prisoners bow to them whenever and wherever the POW came near them. This was also whenever a guard came to their rooms or if they were taken before an interrogator. Those who did not bow to them were severely beaten until they did. It was finally decided among the POWs that it was better to bow to them than to get beaten up each time.

The first few weeks of capture and intensive interrogations were the worst time for a prisoner. He would not know what to expect, and he was all alone, sometimes injured badly. After the initial torture, sometimes the Vietnamese would ease up a little for most of the prisoners. However, if they were pressuring the POW to meet a delegation, make a tape, write a confession or a propaganda letter, or whatever they wanted, they would begin the torture sessions again.

Every few months, the Vietnamese captors would come up with a new idea to use on the POWs. Sometimes it was something like reading news propaganda over the camp radio, writing a detailed biography, or memorizing the camp regulations. Or they would decide to have a communication purge, and many men would be tortured again or put in solitary confinement in very small cells. It was known that if a person gave in to them and started doing what they wanted the way they wanted it without any torture, that person ended up being their go-to, and they would continually be on his back to do things for them.

## After the Music Stops

Attempts at enlightenment were made through history books, magazines, films (rarely), and newspapers. The Vietnamese did not like the term "brainwashing" and called it "enlightening" instead. Ben said that they were given Albanian or Korean magazines, but the magazines were censored. When he would open them up, he would find the pages pasted together. He would peel open the pages to read them and then wet them with his tongue to paste them back together. This reading or viewing material was obviously and comically one-sided and biased. It almost seemed like a joke to the American POWs. The communications were used to educate the prisoners about the history and causes of the North Vietnamese state. It was considered by them to be a real honor, almost a favor, for them to allow the POWs the privilege of reading anything, even communist literature. Ben said he thought to himself at the time that a communist was the only type of person who would torture a person unmercifully to make him sign a statement that he was not tortured. To a citizen of a totalitarian regime or a communist who didn't know freedom, it did not bother them that the pages of magazines or books were pasted together. To an American it was a repulsive thing.

The Vietnamese also gave Ben, as well as others, articles out of Time, Life, and especially Newsweek, but it got to a point where he would look at a picture of a group of demonstrators at a rally in the United States and the police trying to hold them back, and Ben would think that that should not happen in a free country. That's what America is all about, he thought. That's what freedom is. If these peace groups were forbidden from demonstrating, then America was no better than the North Vietnamese and the repressive little society where censorship is not even repulsive. We don't know what suppression is in the United States. We are a free people and stand up for all freedom.

These were posted on the inside of the door to every cell and each POW was ordered to memorize these regulations:

```
      Prison Camp Regulations
    (Imposed by the Vietnamese)

1. The criminals are under an obliga-
tion to give full and clear written
or oral answers to all questions
raised by the camp authorities. All
attempts and tricks intended to evade
answering further questions will be
considered manifestations of obsti-
nacy and antagonism which deserves
strict punishment.

2. The criminals must absolutely
abide by and seriously obey all
orders and instructions from the
Vietnamese officers and guards in
the camp.

3. The criminals must demonstrate
a cautious and polite attitude to
the officers and guards in the camp
and must render greeting when met by
them in a manner already determined
by the camp authorities. When the
```

Vietnamese officers and guards come to the rooms for inspection or when they are required by camp authorities to come to the office room, the criminals must carefully and neatly put on their clothes, stand at attention, bow a greeting and await further orders. They may sit down only when permission is granted.

4. The criminals must maintain silence in the detention rooms and not make any loud noises which can be heard outside. All schemes and attempts to gain information and achieve communication with the criminals living next door by intentionally talking loudly, tapping on walls or by other means will be strictly punished.

5. If any criminal is allowed to ask a question, he is allowed to say softly only the words "bao cao." The guard will report this to the officer in charge.

6. The criminals are not allowed to bring into and keep in their

rooms anything that has not been so approved by the camp authorities.

7. The criminals must keep their rooms clean and must take care of everything given to them by the camp authorities.

8. The criminals must go to bed and arise in accordance with the orders signaled by the gong.

9. During alerts the criminals must take shelter without delay; if no foxhole is available they must go under their beds and lie close to the wall.

10. When a criminal gets sick he must report it to the guard who will notify the medical personnel. The medical personnel will come to see the sick and give him medicine or send him to the hospital if necessary.

11. When allowed outside for any reason, each criminal is expected to walk only in the areas as limited by the guard-in-charge and seriously follow his instruction.

12. Any obstinacy or opposition, violation of the preceding provisions or any scheme or attempt to get out of the detention camp without permission are all punishable. On the other hand, any criminal who strictly obeys the camp regulations and shows his true submission and repentance by his practical acts will be allowed to enjoy the humane treatment he deserves.

13. Anyone so imbued with a sense of preventing violations and who reveals the identity of those who attempt to act in violation of the foregoing provisions will be properly rewarded. However, if a criminal is aware of any violation and deliberately tried to cover it up, he will be strictly punished when this is discovered.

14. In order to assure the proper execution of the regulations, all the criminals in any detention room must be held responsible for any and all violations of the regulation committed in their room.

15. It is forbidden to talk to make any writing on the walls in the bathrooms or communicate with criminals in other bathrooms by any other means.

16. He who escapes or tries to escape from the camp and his accomplices will be seriously punished.

# Eight

Coming down to earth and landing in Montgomery and into reality once our honeymoon was over was a shock. The two children were so happy when we got home, and they had a wonderful summer at their different camps. We were all ready to find a house, get settled, and have a yard for a dog. Then it would be time for school for all of us. And so real life began for our ready-made little family.

We found a nice house on a dead-end street that backed onto a golf course. We moved in and settled enough to begin our search for a dog to make our family complete in our minds. And we found him—an Irish setter puppy we named Ringer who dominated our existence for many days to come. Ben had graduated from the University of Alabama in 1962, so we also were caught up in the fall season of Roll Tide parties, going to the games in friends' homes. Once again, why not join in Ben's celebration of his college's victories and events since he had not been able to do so for so long? The children even got into Roll Tide fever and became little football fanatics just like their stepfather and his brother in faraway Ephesus, Turkey with his Roll Tide baseball cap. Antony and Cleopatra did not have anything over a nutty Alabama fan.

As the year wore on, I noticed Ben getting tipsy at parties and even on off nights when we hit the cocktail hour. I remember thinking that this was normal, as everyone in our circle of friends practiced cocktail hour. When you dropped by to visit someone around 5:00 P.M., you were always offered a drink. It was the norm, and we were still in a celebratory situation.

My mother gave a lovely cocktail dinner at Maxwell Air Force Base Officers' Club and included a lot of the POWs who were attending the Air War College or Command and Staff school at Maxwell. Of course, a lot of Montgomery

## After the Music Stops

people helped us celebrate our marriage. It was a lot of fun, seeing family and friends and fellow POWs. We wore the beautiful evening clothes we had had made in Singapore. My dress was a long sheath with a Mandarin-collared neckline and a slit up the side—an Oriental style—so much fun to wear as well as being exotic.

Ben had stuffed some dollar bills in his shirt pocket. He had paid the babysitter and had money left over, but instead of putting the rest of it into his wallet, being very tipsy, he had stuffed it loosely into his shirt pocket. Jay, about seven years old at the time, had lost a tooth earlier in the day and had announced this fact to all who would hear it. We told him to put the tooth under his pillow for the tooth fairy to find and leave him money there. When we came home that evening, we went in Jay's room to tell him goodnight. Leslie was in her room, and we had already told her goodnight. The next morning, Jay came running into our bedroom, joyfully telling us that the tooth fairy had left over $70, which was spread all over his bed. We gasped because we could not tell him that Ben must have dropped the money. We wondered how much the tooth fairy was going to have to leave in the years to come.

The first year passed with a lot of firsts. Ben's first year of Alabama football season, his first Thanksgiving, first Christmas, and first back-to-school was invigorating for him. He enjoyed the academic atmosphere and spent a lot of time at AUM. One night, he was especially late getting home. I was not too worried at first because he usually stayed late. He said he could study better in the library. It was quieter than our house, with the children running about and the dog barking and the phone ringing—all the usual sounds of a home with growing children living in it. It kept getting later and later, and I began to get nervous because I knew that he often stopped at a bar for a drink or two. I asked myself, "What if he wrecks and is in a ditch bleeding and I am the only one who knows he is missing?" After several hours, I called the highway patrol since AUM was outside the city limits. They notified me after a while that Ben was spotted driving in the direction of our home. After that time, I became indifferent when he was late arriving home at night, unable to walk straight or focus.

In June, our whole family, including the dog, took off for Randolph Air Force Base in San Antonio for Ben to get checked out to fly once more. He was so happy to be in his beloved

F-4 fighter jet again after over seven years. Even though Ben was planning to go to medical school and would not be flying, he wanted to take the Air Force up on its gift of allowing the returning POWs to fly again. This was really just a gift, as most of the men were too old to continue on flight status. It really was a large part of the celebration of these airmen's homecoming. A champagne bottle was opened and poured on the successful pilot at the end of his checkout flight. It allowed the former POWs to know that they still had it, that they were still fighter pilots, and that their country valued them. It was such an exciting event to watch. The children and I celebrated the occasion with a festive dinner and a piñata in the yard, which Ben was allowed to burst, and he shared the candy inside with all of us.

After returning home to Montgomery from San Antonio with a successful completion of flight testing and a champagne celebration, we found admittance letters from the three medical schools Ben had applied to. We decided to take the spot at the University of South Alabama Medical

School in Mobile, Alabama. It was geared toward family practice, which was Ben's preference. Right before our move to Mobile, our daughter Lillian was born.

Also, before the big move, Ben said to me, "I would like to get a second opinion about my neck injury. I can live with it, but it would really be great if it could be fixed or eased at least."

Ben had suffered this injury during his shoot-down. After ejecting from his F-4 and parachuting down, he actually felt the flack from the guns firing at him. When he was captured, the Vietnamese beat him around the neck and shoulders with rifle butts, causing a lot of scar tissue to form around his vocal cords and making it hard for him to talk. When he was rescued years later, Air Force doctors didn't recommend surgery.

We looked all over for specialists that could help Ben, particularly at The University of Alabama's medical school, and we eventually found a well-renowned surgeon that would examine him in Birmingham. However, our hope to better heal Ben's condition were dashed when we were told that the odds of the procedure being effective were low and risky. Ben could be in danger of losing his voice entirely.

# Nine

One evening, Ben began telling me about a rumor they had gotten wind of in prison about a raid of a camp north of Hanoi called Son Tay. It was ironic that he would bring up a particular subject, as I had firsthand knowledge of this raid—another moment when life slapped me in the face. In graduate school, I had met some other students in my thirty-year-old age bracket who were going back to school after serving in the military. They included me in a get-together at their Sergeant Major's home. This sergeant liked to have his group of Green Berets and their friends to his house. He was married to a Vietnamese woman who spoke very little English but seemed to enjoy having all the young people around. Some of these soldiers began discussing a raid they had been on together in Vietnam. I told Ben about what I had learned from this group of men. Special Forces Colonel "Bull" Simons led this raid on the POW camp at Son Tay himself. He recruited 103 volunteers from the 6th and 7th Special Forces Group, and they began studying models of the camp and rehearsing the attack on a full-size replica of the camp. While Simons' men were training, the attack date of November 21–25, 1970 was selected due to having the ideal moonlight and weather conditions for success. After further training, Simons selected 56 Green Berets from his pool of 103. Some of those selected were in this room at the Sergeant Major's house. They spoke with such fervor, and with tears in their eyes, they told the rest of the story.

At 2:18 A.M., the helicopter carrying a fourteen-man assault group landed inside the camp. Others landed outside the camp as support and still more provided security against North Vietnamese reaction forces. The group's mission was to rescue the POWs inside the camp, carry them on their shoulders if necessary, and bring them home. After conducting a thorough search of the camp, it was discovered that there were no POWs inside. Intelligence later revealed that the POWs had been moved fifteen miles away from Son Tay in July.

Even though the Green Berets were dejected that their rescue attempt did not work, Ben said the raid had certainly been good news in the Hanoi Hilton. The POWs who were housed there were elated when they heard about the attempted rescue. The rumor had leaked out because the Vietnamese guards became extremely nervous and feared more raids. They did not want to lose their "bargaining chips" (the POWs) in case they needed to negotiate an end to the war, so after the rescue attempt at Son Tay in November 1970, all of a sudden all the outlying camps prisoners were moved back to the Hanoi Hilton. At the time, there were very crowded conditions—50 POWs in one room. In each of six or seven other rooms were also forty to fifty men.

The rooms were about sixty feet long and about twenty-four feet wide, and for fifty men that was very crowded, but quite an advantage to the American prisoners because they could get better organized. The military structure was already completely organized. It seemed that from that time forward, due to the strength of the POWs numbers, the Vietnamese almost gave up trying to get information or propaganda from them.

These prisoners' average age was 36 and almost all were well-educated officers. Many had gone to the military academies, and others had master's degrees in various areas of expertise. They began teaching each other in college level classes. For example, there were five language classes: French, Russian, Spanish, German, and English grammar and vocabulary. Some of the men took all five languages at one time because they were so hungry for mind usage after years of brain stagnation. The men were teaching these classes without books, pencils, papers, or anything else from the Vietnamese. They made their pens out of bamboo or wire and made ink out of vitamin pills and medicine (which had come in some packages from home that a few POWs had been allowed to receive). The paper they used was made from very rough toilet paper and cigarette wrappers. They had to keep all their provisions hidden from the enemy guards because the Vietnamese prison officials inspected the POWs periodically to find such things. The POWs would clean the floors and the concrete bunk-slabs to use them as blackboards. For chalk, they used pieces of

## After the Music Stops

brick, pieces of tile from the roof, and different colored rocks. At first, the guards tried to stop them from using pieces of tile and brick on the floor, but finally they gave up because every time they would take them away, the POWs would just find some more.

They had to select who was going to teach the classes, which was not an easy decision because there were so many experts to choose from. They chose the one who was the biggest expert in a specific subject. He could be aided by other POWs. The instructor-to-be would spend weeks preparing his lectures, and he sometimes would work up an outline of his course in "chalk" on his bunk and give his lessons from that so he could utilize illustrations and drawings as explanations of the information. They used the term "POW Fact" when something that was proposed as fact was in question when it could not be verified. The POW Fact would have to be checked when they got back home. At least they never gave up hope for their futures, that they would get back home.

The language teachers would write a paragraph of their language down on a portion of the floor each day for their students to try to translate. Some wrote the paragraph in the form of an interesting serial story. Others even combined their artistic ability with their language by drawing pictures to illustrate the paragraph. Some teachers even described movies in their language.

There were also classes in history, political science, biology, astronomy, literature, and poetry. Several men knew literature to the extent that they could recite poems of incredible length and even Shakespeare plays and sonnets word-for-word. These were passed from room to room and memorized by many of the POWs. One of the most interesting classes was in skiing given by a man from Utah. The fifty prisoners in one of the rooms stood on their concrete bunks and tried to do jump turns, stem christies, and hot doggin' while the Vietnamese guards stared curiously at the spectacle. They also had lectures in a less serious vein such as hunting, camping, stereo, photography, book reports, music, and art. Several of the prisoners were very gifted musically. They managed to teach music history, theory, scales, chords, and harmony. One of the music teachers used some of the men as human notes, lining them up to sing the chords and keys. Another teacher would whistle some of the classics and then teach about them.

## After the Music Stops

There were classes in drafting and in art. There was even a class in drawing led by a man who had been an architect, so some of the POWs designed house plans and made three-dimensional projections of them, even coloring them. There was even a Toastmasters club where everyone spoke all about themselves, and it helped them to know each other better. Then they made speeches, telling the men their most embarrassing moment, most hilarious incident, sales talk, and travelogue. There were also debates and panel discussions.

The POWs put on plays, complete with singing commercials, and sometimes they wrote plays of the poems they had memorized with their poetry teachers. The funniest productions were the musicals South Pacific and Sound of Music. One of the most popular prisoners told the story of hundreds of movies, and he knew all the actors in all of them. Another POWs made up his own movies, so when he started telling one, he was told that he could tell the movie every night as long as it was interesting.

The prisoners had American playing cards for a few months, but then they were taken away from them. The Vietnamese took everything from America away from the

## Gloria Gayden Corona

POWs in December 1970, as well as all the toilet articles, games, and more. Then they gave one deck of Vietnamese playing cards to each room. They also gave the prisoners one Russian-made chess set per room. The POWs played chess a little and also played bridge. They had bridge tournaments, and almost everyone in the room played in these. They even gave "master" points and played duplicate bridge.

I learned more about their final days of being POWs at one of the most memorable gatherings of former prisoners in Las Vegas. It was an opportunity for Ben to be reunited with many of his former fellow prisoners who he had not been able to see since his liberation from Vietnam. I looked forward to meeting the ones he had told me about. Their joy at seeing each other again and the affection that they displayed toward each other was quite moving to observe. This Red River Rats group was named after the Red River in Vietnam that these American pilots had flown over. It is the largest river in North Vietnam, flowing from Yunnan in Southwest China through Northern Vietnam to the Gulf of Tonkin. The pilots who flew the Red River Valley in North Vietnam were known as River Rats, but the enemy there called all American pilots in Vietnam "Yankee Air Pirates."

## After the Music Stops

The first official Red River Valley Fighter Pilots Association reunion was held in 1973 after the release of the POWs the previous spring.

This is their mission statement:

> "We are an organization of military aircrew members. We are dedicated to providing a college education for the children of all military aviators killed in the performance of aircrew duties. Our origins are from the Vietnam War, and the name comes from the Red River that runs through Hanoi. Charter members are those that flew combat missions over the target areas. All of the Vietnam POWs are members of our organization, including a few Medal of Honor recipients."

One of Ben's friends from prison was Lieutenant Colonel Jay Jensen from Utah, and they were so happy to be reunited. It was the first time they had seen each other since coming home. All of Jay's repatriation took place at March Air Force Base in California. I had heard Ben talk about being one of the reluctant twenty, and Jay was also one of them. Ben and Jay explained to me that on January 30, 1973, the peace agreements were read in the Hanoi Hilton to the POWs, and a copy was given to them. The sick and injured were to go home first. Then all others would be sent home in the order of who was captured first after their shoot-down, in groups of about one hundred. On February 12, 1973, the first 112 POWs were released. The next release would be in about two weeks. Then, unexpectedly, an earlier release was announced for 20 names. They were informed that, as a tribute on Dr. Kissinger, 20 extra prisoners would be released early and not in order of shoot-down date, which was extremely suspicious to the POWs. The Vietnamese said that the United States government had requested them by name. The prisoners did not believe that and demanded proof. Finally it was admitted that the United States had requested 20 POWS, but not by name, and this really made the Americans angry. They told the Vietnamese they were playing games, lying, trying to make the POWs look bad. The men said that they would absolutely not go home in this manner and that they would wait their turn. That was very humiliating to the North Vietnamese, as it caused them to lose face. The next day, the Vietnamese told the POWs that the release was already arranged with the United States government, and their names had been given to the press, but the prisoners said that that was too bad. They were still not going. The day after that, they were moved back with the rest of the POWs. The American soldiers told their commander the story and were informed that the return of the rest of the POWs would be delayed if they did not go. The commander gave them a direct order to go, so they were finally released on February 18, 1973.

# Ten

After graduating from medical school, it was time for Ben to go into a residency program. He began to send applications and travel around the country to interview at teaching hospitals for a program in family practice. The children were interested in some of the openings, as they thought it might be fun to have a change of scenery.

In one way, it seemed it might be a fresh start at a new place, but on the other hand, it was a slightly scary prospect. Ben was drinking so heavily at that point that he was out of it most nights. Life was so busy with three children; however, it failed to make a deep impression on me. It seemed it was one more thing to compartmentalize and face another day. I am sure my preoccupation with the children was a convenient distraction to avoid facing Ben's drinking and all that his drinking would do to my little family. I had already been through one divorce and couldn't bear the thought of another one. A separation would also mean that after all the sacrifices, he couldn't keep a medical license and all his dreams would be over. My dreams would be over too. As a doctor's daughter, I knew what an honorable profession medicine was and had been willing to make the sacrifices to help Ben reach his goal.

The scary part was moving far away from family and friends with my three children in tow. What if Ben became unhinged and needed hospitalization or, worse still, ran out of money and had to get another job or move to Alabama with no help of any kind? After a lot of agonizing days and nights, Ben settled on a hospital in McAllen, Texas. He said it would provide a really good training situation, as the hospital was used by patients of all ages and races, which would let him experience a wide variety of illnesses. Ben and I flew down to this town on the southernmost border of the United States and were charmed by the area. Ben and I both liked the Rio Grande Valley for its Hispanic influence as well as the cowboy environment. We checked out

## After the Music Stops

the schools and even found a house not too far from the hospital with a nice yard for our new dog, a collie named Jingle Bells.

We made the move to Texas, and Ben began his residency. The children and I spent our days settling in and getting to know our neighborhood. Our next-door neighbors turned out to be our dearest friends. Nancy Fox was a single mother with a child, Toby, who was Lillian's age. She had a backyard swimming pool that my children loved to spend their day splashing in. Ben seemed pleased with his program, and for a while our lives were most pleasant. He had a study in the converted garage of the house with its own entrance. It seemed ideal, as the house was always "rocking" with three active children and a rambunctious collie. When the big children were in school, there was still four-year-old Lillie doing her ballet moves all around the house. Ben always complained about the noise and said that he needed an alcohol-infused drink to calm his nerves. I must say that it was difficult to keep a quiet home, so his separate studio was really a plus for our family life.

Several months into his residency, Ben quit. He said that the other two residents colluded to leave him stuck

## Gloria Gayden Corona

alone covering the family practice portion of the hospital service. He felt that they were teaming up on him and that he did not feel he was ready to cover the needs of the patients. He said he was afraid if he reported these residents to the director of the program, these senior men would have it in for him. I detected possible paranoia but I dared not mention it to him. I felt that he needed my support, not my criticism.

So the family was on the move again. It took many months for Ben to find a new residency to accept him, as all the slots were already filled. In the end, a program in Alabama accepted him. When I was packing up the household goods in preparation for the moving van, I went out into his study to put all of our books into cartons and found empty whiskey bottles hidden everywhere—behind loose bricks, in desk drawers, under furniture. I was beginning to grasp the real severity of the problem.

Once Ben finished his residency program in Montgomery, he went into a joint family practice with another doctor in Mobile. As time moved on, he became paranoid that the doctor he was in practice with was cheating him on the payments from patients. He became more and more emotionally distraught and drank even more. When I spoke to him about his troubling behavior, he said he had to self-medicate to deal with his ghosts and demons. I think I had already guessed that Ben was drinking to excess in order to drown out his imprisonment in Vietnam. But when he confessed this to me, I felt it was a good step for him to share his pain. He must have known that I admired him for his bravery or I would not have been married to him. However, I could already see the alcohol affecting his health. His skin was sallow, his eyes were bloodshot, he was extremely thin, and his stomach was distending. It also seemed like his hands were perpetually shaking.

At this juncture in our lives, we needed to explore our options for Ben's medical practice, as we had three children to support. It would be expensive to go into practice alone because of all the equipment he would need. About this time, Ben was approached by a banker friend from Elba about him coming back to Elba to practice medicine. It was an offer too good to pass up. The office would be paid for by the city since they did not have a doctor in town—the nearest doctor was miles away. The city would furnish the office with all the equipment that a physician needed for a family practice, and they would hire and furnish the staff. They would pay Ben a salary, and therefore all Ben would have to do was show up and attend to patients. There were some problems that would have to be addressed though, such as where we would live and the older children's schooling situation. Our oldest was close to graduating from high school and would be leaving home for college soon, so she would not be affected much by the move. The second child, however, was a serious and dedicated student, enrolled in a stronger, academically focused school in Mobile, and was aiming for the Ivy League and a scholarship. We knew his ambitions and wished to further his preparation for his goals, so we agreed to send him to a private boarding school in Tennessee, as we knew he probably would lose his momentum in the Elba schools. The small Southern schools in rural areas of Alabama did not, at that time, offer the advanced placement classes that were needed to be accepted at high academic universities. I had to ask Ben to guarantee this expense for Jay because the move to Elba would cost me dearly in the loss of my occupation, plus Ben was becoming more and more unstable in his drinking. I did not want to gamble with my son's future any further.

## After the Music Stops

We agreed to take this offer and began the process of relocating. We put our house on the market and found a home in Elba. Since the house in Mobile had not yet sold by the time the Elba office was ready to open, Ben had to go on ahead of the family in the move. He drove the five hours to Elba and stopped by the new office that was to open for patients the next morning. He went in and noted the gleaming waiting room full of flowers and plants from well-wishers to him, the hometown hero.

He went to a motel and called me. He said he could not go through with this move, that he had "hit a wall" and could not continue. He said that all of his childhood trauma, his restrictive parents, and his anger at them just welled up in him. And then there were the Vietnamese and their mistreatment of him and his intense anger toward them. Needless to say, I was horrified at his revelation and suggested he sleep on it. The next day, he returned home to Mobile and spent several days sleeping as if he were terribly ill. From there, we had no alternative but to go into debt and open a solo practice in Mobile.

# Eleven

After the move to Elba was cancelled and we were in the process of opening Ben's medical practice in Mobile, we began to settle into a normal-on-the-surface existence all together back at home. That is the only way to describe our daily life. One would observe our family and think that we were pretty much just like every other one there. In our life, each family member went about his or her own daily routine and schedule, separate from the others. When we came together at the end of the day, the house had a hush about it, the children retiring to their own rooms and staying there so as not to chance an encounter with a drunken Ben, who would pick a verbal argument with whomever he came face-to-face with, always bringing about some heated confrontation.

It was slow-going setting up of a solo medical practice—selecting office space, buying medical equipment, decorating, and furnishing the waiting room and examining rooms. Ben's office needed a desk, frames for all of his licenses, and decorative pictures for the office space. There was so much to do that I could not take a breath in order to assess the situation unfolding around me. Since I would manage the office instead of having to hire an office manager, it would save money in the beginning. I knew that Ben was drinking heavily at night, but I thought that that was due to the stress of our situation. The children were busy with their lives and happy that we were not moving away and leaving their friends.

On a weekday, while I was getting ready to leave for the office, a close friend of ours, Tom, called and wanted to talk with me privately. I met with him, and he accused me of being like "an ostrich with my head in the sand" about Ben's drinking. It seemed that many friends had noticed his alcohol consumption, his slurred speech, and his obvious poor health. This friend wanted to set up an intervention with

## After the Music Stops

some of Ben's family and closest friends and see if we could get Ben into treatment. I had noticed his short temper and irritability, and when I had mentioned his drinking, he had become so belligerent that I became afraid of him physically. He would have violent outbursts at night and be so sorry the next day, sending flowers as an apology. Sometimes he seemed to black out, not remembering the actions of the night before. On several occasions, he would claim that he sensed something had happened the night before, but he could not remember what it was.

One incident, one that is both comedic and tragic, comes back to me quite clearly. Ben and I had been invited to an evening cocktail party by a couple who were close friends of ours. The doctor host had been a flight surgeon in the Air Force, and he had shown an interest in Ben's condition, particularly in the way he medicated himself with alcohol. I hired a sitter for Lillian because the two older children had their own engagements.

Ben and I drove together to the party, but as the evening wore on and I became tired, I drove home alone. Ben seemed to be having fun talking to a lot of doctors, and the alcohol was flowing. I spoke to our host and asked him to see that Ben got home from the party safely. He was happy to do so. Since it was in the early morning hours, I had needed to let the sitter at our house leave. I waited up for a while but finally gave up, thinking that Ben would wake me up when he got home.

In the meantime, Ben slipped away from the party unnoticed and started walking home down the quiet, deserted residential streets. Seeing a strange, obviously drunk, tottering man, the people on either side of the streets observing from their windows became fearful and called the police. Ben was ultimately picked up and put in the back of the patrol car. The two police officers rode around the city with Ben in the rear seat until the sun came up. They then knocked on our front door and informed me that they had my husband in their custody. I asked them why they had not put him in their downtown drunk tank, to which they replied that they did not put doctors there.

I agreed to do the intervention, even though I knew it was going to be extremely difficult. Tom contacted Ben's brother in Birmingham, and they set up the date and people to participate. His brother booked a room in a treatment facility in Birmingham, which had a separate wing for doctors "drying out" so their patients would not know and the doctors would not lose their livelihood. We knew that this would eliminate one of Ben's reservations about going into treatment. He had told Tom and me on separate occasions that he used alcohol medicinally in order to cope. It was later decided that I would not participate in the direct confrontation of the intervention to keep Ben from becoming hostile to me. They had the intervention, and Ben agreed to go into treatment. There was a three-day gap between the actual intervention and the availability of his room in the Birmingham hospital. On the third day, Ben said he could not go, that he could not face being locked up again. After I said to his face that I knew he was an alcoholic and I told our minister and some friends leading to the intervention, Ben saw me as the enemy.

Since I had become Ben's new enemy, just after the Vietnamese and the United States government, Ben moved out of the house. The children, as well as the house itself, breathed an audible sigh of relief. We all had been tiptoeing around an irritable drunk person every night, and it made the atmosphere in our home quite unpleasant, so much so, that the children quit bringing their friends over to our home. They were embarrassed by his behavior and afraid of what he would do. Ben was a mean drunk and would taunt people and make fun of them with sarcastic remarks.

At first Ben, moved into a motel on Government Street in Mobile and spent the nights there in the motel lounge-bar. In the daytime, he would go to his office to see patients. One evening, he knocked on the door to the house and asked to see Lillian. He said he needed to give her a hug and tell her goodbye, as he was going to have to go into hiding. He said the FBI or the CIA was after him, and he did not know why. He said he could see their cars drive by the motel and turn their lights off as they drove by. As this motel was at a busy intersection, it defied logic that anyone would turn their lights off and think they could drive by undetected on a much traveled street. I allowed him to see Lillian for a few minutes on the condition that he not tell her any of this. She was too young to have contact with her father's paranoia.

The frequency of these types of incidents increased until the Veterans Administration declared that Ben suffered from PTSD and he began treatment in a VA hospital. As the years passed, Ben was in and out of treatment in VA hospitals and clinics, and in the end, he was forced to give up his medical practice. I have often been asked what else happened to Ben after our divorce. Truthfully, I don't really know what all happened to him. It's been rumored that he moved several times, married again several times, and ultimately lost his medical license, but again, this is all hearsay on my part.

In 1998, he was found dead after what appeared to be about a week, sitting alone on his sofa, from an apparent heart attack. Ben was buried at Arlington National Cemetery with a caisson pulled by six white horses and with full honors, his casket adorned with an American flag that had been presented to Lillian. While what happened in his final moments may be disputed, and despite the illness that eventually took its toll on him, there is no doubt that Ben Ringsdorf died a national hero.

— Gloria Gayden Corona —

## Veteran Tributes

*"Honoring Those Who Served"*

Ben Ringsdorf was born on September 16, 1939 in Dothan, Alabama. After completing his bachelor's degree, he entered Officer Training School with the United States Air Force on November 9, 1962, and was commissioned a 2nd Lt. at Lackland Air Force Base, Texas on February 5, 1963. Lt. Ringsdorf next completed Undergraduate Pilot Training and was awarded his pilot wings at Vance Air Force Base, Oklahoma in March 1964, followed by Pilot Systems Operator Training and F-4 Phantom II Combat Crew Training from May 1964 to February 1965. His first assignment was as an F-4 pilot with the 47th Tactical Fighter Squadron at Mac Dill Air Force Base, Florida from March 1965 until he was forced to eject over North Vietnam and was taken as a Prisoner of War while flying on temporary duty with the 559th Tactical Fighter Squadron at Cam Ranh Bay AB, South Vietnam on November 11, 1966. (This was his 3rd tour of duty in Vietnam). During his time with the 47th Tactical Fighter Squadron, Lt. Ringsdorf *also flew combat in Southeast Asia from July to November 1965. After spending 2,286 days in captivity, Capt. Ringsdorf was released during Operation Homecoming on February 12, 1973. He was briefly hospitalized to recover from his injuries at Maxwell Air Force Base, Alabama, and then attended pilot recurrency training at Randolph Air Force Base, Texas before leaving active duty on October 9, 1973. After leaving the Air Force, Ben completed Medical School and worked as a Family Physician until his death on February 15, 1998. He was buried at Arlington National Cemetery. (His daughter Lillian received the American flag at his burial.)daughter Lillian received the American flag at his burial.)*

# Gloria Gayden Corona

## Silver Star Citation reads...

*For gallantry in connection with military operations against an opposing armed force as an F-4C Pilot of the 559th Tactical Fighter Squadron, 12th Tactical Fighter Wing, Cam Ranh Bay, Vietnam, in action in southeast Asia, on 11 November, 1966. On that date, Lt. Ringsdorf piloted an F-4C aircraft against a target in North Vietnam. Despite known defensive weapons fire reported by other aircraft in the attacking flight, Lt. Ringsdorf selflessly continued his attack. With complete disregard for his own personal safety, Lt. Ringsdorf made low repeated passes over suspected anti-aircraft gun positions to draw their fire and expose themselves to attacks from fighter aircraft. By his gallantry and devotion to duty, Lt. Ringsdorf had reflected great credit upon himself and the United States Air Force.*

― After the Music Stops ―

## We Came Home
*1977*

Captain Ringsdorf was shot down in his F-4C aircraft on a combat mission over North Vietnam on November 11, 1966. He joined the Air Force in 1963 shortly after graduating from the University of Alabama with a degree in Chemistry. Until his return from North Vietnam in 1973 he had remained a bachelor. Then Governor and Mrs. Wallace introduced him to Gloria Gayden. On May 3, 1973 they were married in the Maxwell Air Force Base Chapel, Montgomery, Alabama with Governor Wallace and his wife, Cornellia Wallace, as members of the wedding party. On February 1998, Dr. Herbert Ringsdorf resided in Alabama until his death from a heart attack. He is survived by his daughter, Lillian Macon Ringsdorf.

Ben Ringsdorf's medals include: the Silver Star, 3 Distinguished Flying Crosses, Bronze Star, Legion of Merit, 2 Purple Hearts, Air Medal, Presidential Unit Citation, Air Force Outstanding Unit Award, Vietnam Service Medal and 1 Prisoner of War Medal.

# Epilogue

When I have discussed the subject of this book, I am constantly shocked by the reaction of ordinary Americans of various ages. They have no idea of the background of the Vietnam War or even why the war was fought. I have researched the subject and feel obligated to share a little about its history.

The Vietnamese have fought the Chinese for over 1,000 years, and although they were occupied and ruled by China during much of this time, they were never conquered or defeated totally. Over the 1,000 years, there were always a Northern part, or a principality, and a separate Southern part. China coveted Vietnam because of the rich, abundant rice-growing resource. It is understood by scholars of the area that the Vietnamese fear China's dominance more than anyone else, and they are careful in their dealing with China.

As early as 1787, the French colonized Vietnam and dominated all of Indochina. In 1940, when France surrendered to the Germans, Indochina came under Japanese control, who left the control nominally in the hands of the Vichy regime of France. In March 1945, the Japanese interned the French Vichy administration and troops in Vietnam and proclaimed the State of Vietnam was under the rule Bao Dai, emperor of Annam. This puppet state collapsed after the surrender of Japan later in 1945. Since France was not in a position after World War II to reoccupy Indochina, it was agreed at the Potsdam Conference that the Nationalist Chinese would occupy north of the 16th parallel and British troops, south of it until the French could resume control. However, in August 1945, the Viet Minh League, led by communist Ho Chi Minh, proclaimed Vietnam as the Democratic Republic of Vietnam. France reestablished itself in Vietnam and England, and China withdrew. Ho Chi Minh headed a war of independence against the French and was victorious in May 1954. A provisional military demarcation line along the 17th parallel was agreed to with the territory above the line going to the Viet Minh and south of the line going to the Saigon government, led by Bao Dai. Under the Geneva agreement, elections were to be held in 1956 for a unified government in Vietnam. In 1955, however, Premier Diem sponsored a referendum in South Vietnam ousting Bao Dai as Chief of State. He proclaimed South Vietnam a republic, proclaimed himself president, and refused any election talks with North Vietnam.

In 1954, after the cease-fires, over 200,000 Vietnamese Catholics went to South Vietnam to live. Most religious people in the 1960s were Buddhists, and more than two million were Catholic. Most of the Catholics at the time of the Vietnam War were in South Vietnam. All anti-communists, shopkeepers, teachers, and others went south also.

The return of the Vietnam POWs in 1973 seemed like a morality play carried out by the media. It was something joyous coming out of the disaster in Indochina, something to unify a nation divided by war.

Whether or not a person was for or against the Vietnam War should not affect what a person felt about the POWs, in my opinion. Some of the pilots who were shot down were not necessarily "for" the war, but they supported their country and their fellow citizens. I was proud of Ben's service to his country and his bravery in the line of duty. To survive the prison camps in North Vietnam where he spent six years and six months, he had to have great strength of character. The POWs I met after they returned home were the best of our best in America. I feel privileged to have known them all.

The media portrayed the returning POWs as a band of heroes, which they were not. Anti-war activists insisted they were criminals and liars, which they were not. The Haydens, Fondas, and Clarks were proved foolish. The most thoughtful comments were made by a reporter from the Midwest named Michael Miner, who also was a Vietnam veteran.

"The POWs are too important for us to slough them off so cheaply—and that is what we have done, with stories that were old the second time we wrote them. I want to know just who the hell these prisoners are, because some part of them is part of me and some part isn't, and the same overlap is true of the country. The country must know what to make of them and at the moment it doesn't, just as we reporters don't; yet their return mattered.

## After the Music Stops

It was more than just another cheap whiff of sentiment; when they came back; an era ended."

# Afterword

It has been reported that presently, even to this day, veterans have been found in alleyways, alone, lying dead in their own filth with an empty whiskey bottle close by. For one such person, it was obvious that he felt that he had nothing to live for and that the force of post-traumatic stress disorder had him in its grip. This observation can be made of an extremely high number of returning soldiers from present day wars. This book documents that this fact can also be true of returning men who have everything to live for. The veteran highlighted in this story was one of the aforementioned ones, with everything going for him, but he still was found dead, alone, after about a week, with empty whiskey bottles around.

— Gloria Gayden Corona
Fairhope, Alabama

www.ingramcontent.com/pod-product-compliance
Lightning Source LLC
Chambersburg PA
CBHW070620300426
44113CB00010B/1601